DESIGNING WITH DRIED FLOWERS

DESIGNING WITH DRIED FLOWERS

Creating Everlasting Arrangements

HANNAH ROSE
RIVERS MULLER

PHOTOGRAPHS BY
MOLLY DECOUDREAUX

CLARKSON POTTER/PUBLISHERS
New York

Published in the United States by Clarkson Potter/
Publishers, an imprint of the Crown Publishing Group, a
division of Penguin Random House LLC, New York.
ClarksonPotter.com

CLARKSON POTTER is a trademark and POTTER
with colophon is a registered trademark of Penguin
Random House LLC.

Library of Congress Cataloging-in-Publication Data is on
file with the publisher.

ISBN: 978-0-593-58098-1
Ebook ISBN: 978-0-593-58099-8

Printed in China

Editor: Deanne Katz
Production Editor: Ashley Pierce
Designer: Mia Johnson
Production manager: Kim Tyner
Compositors: Merri Ann Morrell and Zoe Tokushige
Copyeditor: Erin Cusick
Proofreader: Andrea C. Peabbles
Photo retoucher: Molly DeCoudreaux
Publicist: Lauren Chung
Marketer: Monica Stanton

10 9 8 7 6 5 4 3 2

First Edition

*To my mama,
the most important and
magnificent flower
in my life.*

Introduction

My mom and I speak to each other through the flowers we grow. The joys and triumphs of our flower fields and bouquets are etched into the smile lines around our eyes, and each late frost that hit our spring flowers or gopher that found our tulips has added a wrinkle to our furrowed brows. We gawk over seed catalogs together, wondering whether new flower varieties would fare well in our growing zone. We harvest together early in the morning. We dream the same dreams of snapdragon fields, mixed bouquets, and fragrant wreaths.

My mom, Dru Rivers, has been growing flowers on our farm since 1984. In the beginning, Full Belly Farm grew mostly organic fruits, vegetables, and nuts, but my mom believed that organic flowers were equally important to the sustainable food and local farm systems in the United States. She saw what it meant to care for the pollinators by planting flowers. She enjoyed seeing the transition from seed to bloom and loved being able to pass on her enthusiasm for the natural world to her children. So she experimented with flowers until she knew how and when to plant each seed and learned what grew well in our dry, hot climate. She would bunch sweet peas on our kitchen floor after her four kids had gone to bed so that the flowers would be ready for market and orders the next morning. Slowly her passion for growing flowers grew into more than just a hobby, and her love for flowers was contagious. She ignited a desire for locally grown flowers in all the stores to which Full Belly

Farm sold, and, as the years went on, the farmers' market customers begged for more. So my mom started growing on an acre plot of land, increasing slowly year after year. As her babies grew, so did her small flower-farming operation.

I was always uniquely intertwined with the flowers that my mom grew. I slept in harvest boxes as she weeded and harvested out in the flower fields. I spent the earliest parts of my childhood roaming through her garden, watching her experiment with growing flowers, learn to trellis rambling sweet peas, and dry her own flowers; she began to imprint a deep love for flowers and farming into my hands and heart. I built fairy houses and foraged bouquets and flower crowns with the flora I found on my adventures around our farm.

Over the years, the small, old garage that had belonged to the previous farmer of the land became the room where my mom experimented with drying extra flowers for wreaths and bouquets for the fall and winter. This was my favorite place to spend afternoons. I would gaze in awe at the colors, textures, and smells that hung from the ceiling's beams. I made my first wreath when I was about four. I learned the techniques by watching my mom and her friends work to make wreaths for the markets where they sold their flowers and produce. I practiced for years and years, and even though my first wreaths were lumpy and misshapen, my mom hung them with pride in our house—just as my grandma had done with my mom's first wreaths.

My love for flowers has endured. I moved away for college and eventually moved back to our farm to begin my floral business. My adoration has not wavered even as my relationship with my mom changed from simply mother-daughter to best friends and co-workers. And now, after years of working with flowers in the field, arranging fresh flowers for countless weddings and events, and learning the art of dried flowers through trial and error and lessons passed down from friends and family, I hope that I can impart some of my love and knowledge of flowers to you.

Within the pages of this book, you'll find stories, recipes, and inspiration from a year of growing, drying, and arranging with dried flowers at Full Belly Farm. We have compiled two generations (my mother's and my own) of knowledge on which flowers dry best, when to harvest each bloom to receive the best outcome, and what to create once they are dried. I explain the tools and foundation that you will need to create your own space to dry flowers, and I share lessons we have learned along the way. The third part of this book is devoted to all of the amazing things that you can create with dried flowers. You'll find small arrangements for every day, beautiful wreaths for every season and occasion, and celebration showstoppers that will look beautiful in the moment and for months to come.

This book is meant to instruct, inspire, and encourage you to dry your own flowers and begin to live with them intrinsically.

Dried flowers have a way of connecting us to the past through sights and smells that remind us of the special moments of our lives while simultaneously allowing us to create something that will last far into the future. Overall, I hope to impart even a small amount of what my mother gave to me: a lasting love for flowers.

FULL BELLY FARM

As I drive down the road to the heart of Full Belly Farm on my way to work in the morning, the sun is just peeking over the Capay Valley hills. The light creates a warm glow across the valley, kissing the tops of the walnut and almond trees and finding its way down to the rows and furrows across the valley floor. I notice the sheep have been moved from the day before, painstakingly herded to greener fields, and the chickens have been let out after a night tucked into their roosts of their transportable chicken coop. The Great Pyrenees farm dogs, who herd and shepherd the animals and children on the farm, stop to sniff each car and truck that is parked along the main road that leads to the center of the farm.

More than fifty people make their way toward the shop and office, as they do six days a week. Families, friends, and co-workers greet each other in their morning ritual as they move into their workday. Many of the friendly and familiar faces I see each morning have been working with the farm for longer than I have been alive. Celso whistles as he loads boxes onto the back of his rusty work truck and I take my (albeit slightly cold) cup of coffee and walk toward the flower shed, where the eight women with whom I work will discuss the day, divvy up tasks, look at the pick list of orders, and talk about flowers that need to be harvested and hung for drying.

The flowers at Full Belly Farm are just a small piece in a complex and ever-evolving operation that employs more than seventy people; grows more than three hundred different types of produce, grains, nuts, fruit, flowers, and animals; and hosts countless farm dinners and weddings. When my parents started Full Belly Farm back in the '80s, they believed that diversity was the key to the health of the soil, to a small farming business, and to the future of their farm. They took that word, "diversity," and ran with it for the last forty years. At any given time, you can find people loading trucks, harvesting zinnias, making tomato sauce, scheduling our Community Supported Agriculture (CSA) boxes, picking plums, harvesting grain, washing eggs, fixing equipment, spreading compost . . . the list of activities and jobs and the stories that could be told about Full Belly Farm are endless.

Flowers may be just a small part of the farm puzzle, but they are my whole world. The farm's flower enterprise can be broken down into three main sections: Filling orders for fresh flowers, which we sell through wholesale, stores, markets, and our CSA boxes; preparing flowers for weddings and event design; and harvesting flowers for drying and creating dried flower arrangements.

FARM
KITCHEN
FULL BELLY
Farm
Certified Organic
OFFICE &
DELIVERIES
MARKET

We spend our days harvesting in the mornings, filling orders, and bunching bouquets for markets and stores. We give our CSA customers a beautiful mix of whatever is blooming most prolifically in the fields at the moment as well as arrange bridal bouquets, construct arches, and fill buckets of blooms for DIY orders. We dry what we cannot sell fresh and also grow things specifically for drying. In all, about half of what we grow fresh gets harvested, hung, and dried in the Wreath Room.

Then, in the fall, once the first frost hits, the colorful flower field turns brown, and our attention turns to the flowers that we, like little field mice, have been hoarding away for the past year. We pull them out of their boxes and begin to create dried flower arrangements. We live intimately with the dried flowers. We incorporate them into gifts, tablescapes, wreaths, bouquets, and home décor. As busy farmers, we can easily forget to bring fresh flowers into the home or get frustrated with how quickly fresh blooms die on our kitchen tables. That's why dried flowers are farmers' preferred flowers, and I know they will be for you as well.

THE WREATH ROOM

The Wreath Room at Full Belly Farm is at the heart of countless stories. While the main purpose of the room is to dry flowers and make wreaths, hence the namesake, it's also been home to countless Thanksgiving dinners, classes, and craft parties, and has even been a sleeping place for litters of kittens and chicks. The room itself is worn; the woods of the workbenches and tables are smooth from years of activities and projects. The soft terra-cotta floor was hand-tiled by my mom, replacing an era of green shag carpeting. The space is the perfect combination of functional and ethereal that has come from years of fine-tuning and trying to keep an order to the flowers that we box up and store away for the fall.

The Wreath Room is thick with the perfume of more than fifty varieties of flowers and a history of more than four decades as the drying room for Full Belly Farm. The shed itself was built in the 1930s by a previous family who inhabited and farmed the land before my parents purchased it in the 1980s. Anyone who enters will see a room full of hard work and passion in the hanging flowers, but the years of stories are harder to see and require keen observation. There's the blackened part of the roof, where the barn caught fire and almost burned to the ground late one night. And the wooden shelves that now hold saved seeds of favorite flowers and dried flower crowns and bouquets from weddings. There's evidence that the room was originally built to be a mechanic's workshop—the wall still opens on the west side by a rickety, old, manual garage door.

The Wreath Room is only about fifteen by forty feet, and the simple drying methods are on display in the sixty or so wires strung tightly between the beams of the building and the nails hammered haphazardly into the walls. An old stepladder on wheels makes hanging and taking down bunches easy. Sheets of labels, a trash can and compost bin, and rubber bands make everything simple and straightforward.

In the summer, harvest crates come in from the fields each day filled with flowers that need to be hung. We often don't have time to clean, so the floor becomes riddled with seeds of every kind, and half-empty cardboard boxes sit in the corners of the room waiting to be filled to the top and labeled. But when I do sneak away from other midday tasks and finally find time to tidy up, sweep the floors, and wipe down the windows, the light shines through the west-facing window and the floor shines in the afternoon light. All of the newly hung flowers show their bright faces, the Wreath Room sparkles its dusty sparkle, the bustle of the outside world fades away, and I remember why this is my favorite place in the world.

SOURCING AND DRYING YOUR OWN FLOWERS

Whether you are simply looking for dried flowers for a final project or you are interested in experimenting with drying the flowers yourself, this section will guide you in that journey. Sourcing flowers that have already been dried may be best if you have limited time or space or if you want a specific flower you cannot grow yourself. But, if you are ready to take on the exciting challenge of trying to dry your own flowers, there are many ways to go about it. You can purchase fresh flowers that you know will dry well, forage for grasses and foliage in your own garden or open spaces that allow for harvesting, or grow your very own!

Whether you work with pre-dried flowers or fresh flowers that you dry yourself, try to source the freshest product possible. This will ensure you get the longest-lasting vibrant colors, least damaged stems and flowers, and products that are naturally treated and dried.

A NOTE ABOUT BLEACHED FLOWERS

There are a number of ways that commercial dried flowers are preserved. Some of these techniques involve harmful chemicals, bleaches, dyes, and preservatives. While this is a trend I have seen take off in recent years, we at Full Belly Farm—as an organic farm and a place that believes in creating a more sustainable planet for future generations—have purposely opted out of such preservation techniques. Bleached and dyed flowers cannot be composted, which means they must go in the landfill, creating unnecessary waste.

If you want to naturally bleach your own flowers, all you need is sunlight and time. Often, it's preferable to hang your flowers in dark areas to preserve their rich color, but hanging them outside or in a sunlit room for a longer period can achieve the same neutral look as chemically dyed or bleached flowers without the harmful substances.

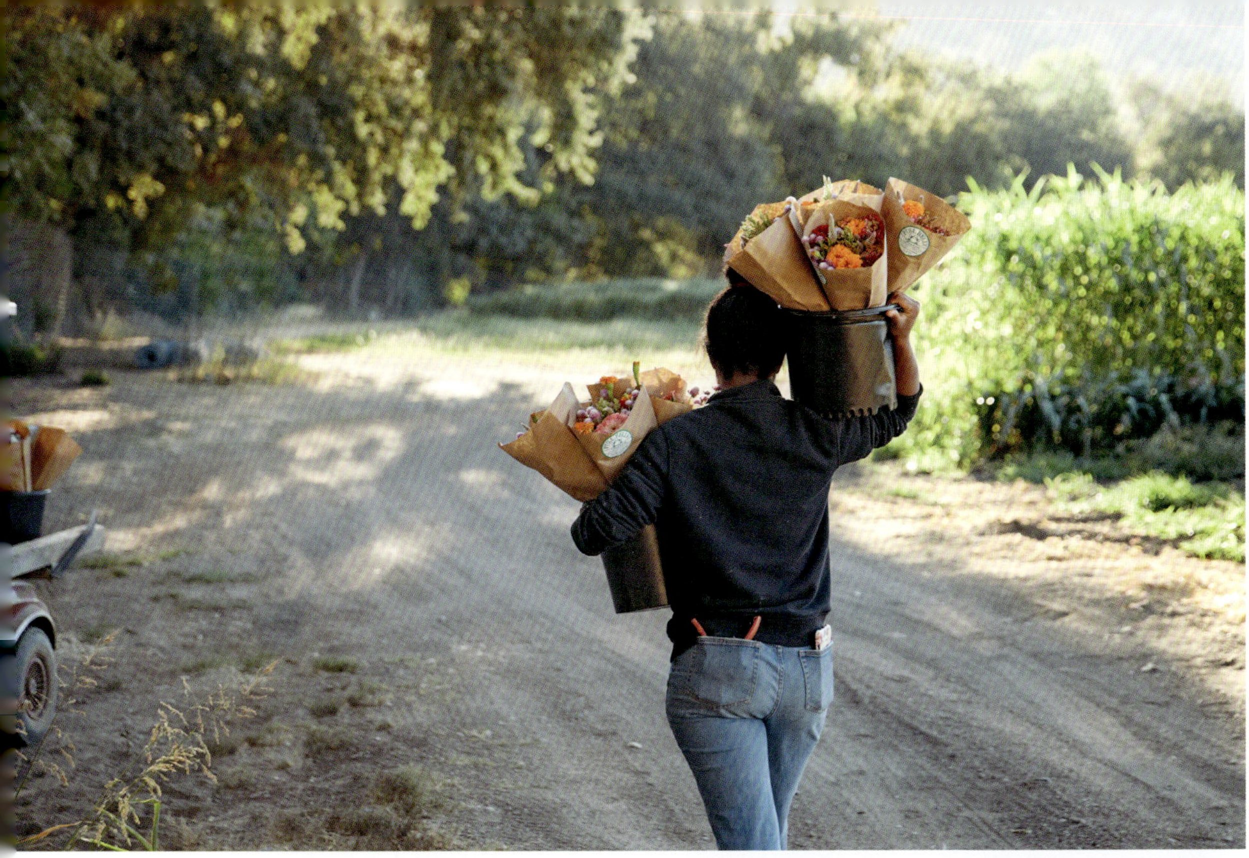

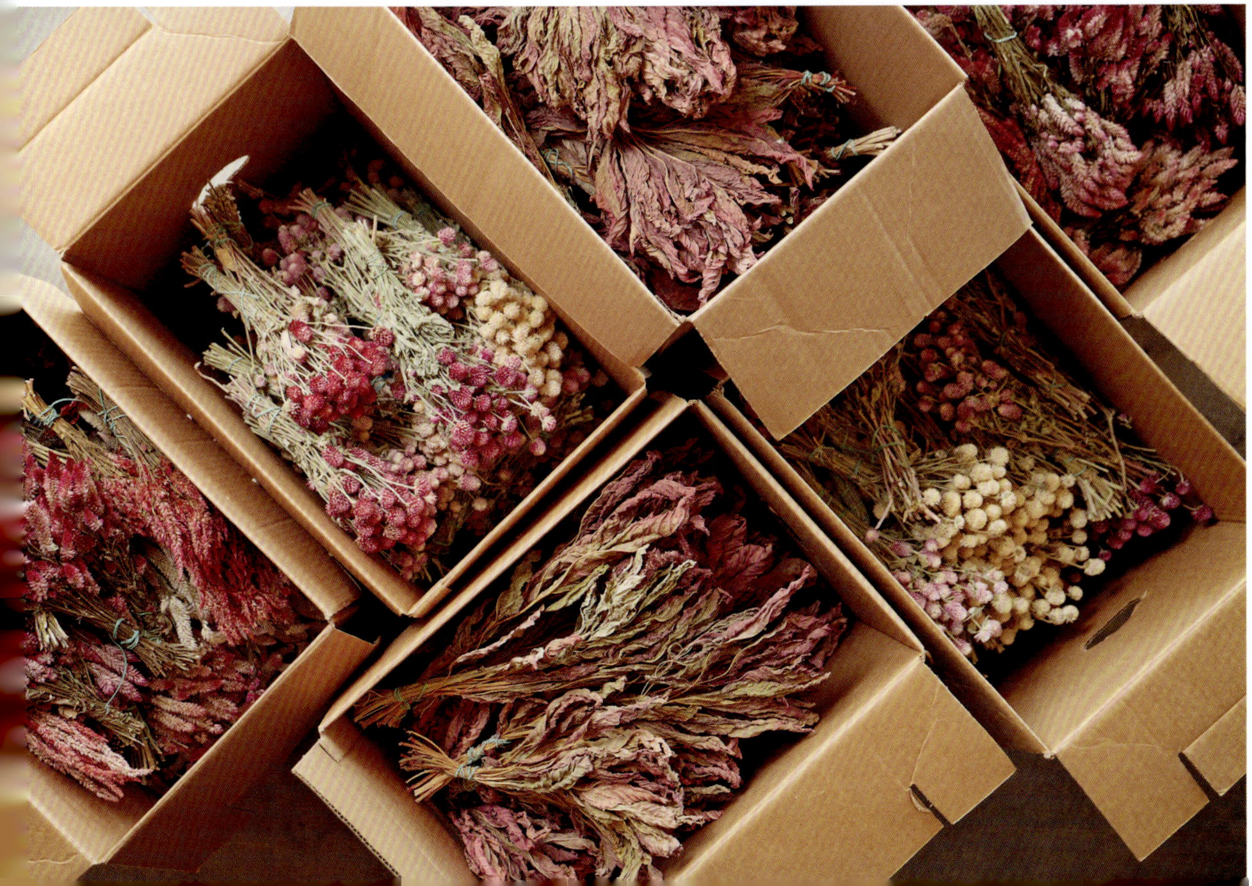

SOURCING

If you want to get started right away arranging with dried flowers and can't imagine waiting for your own flowers to dry, there are many ways to purchase flowers that are pre-dried. Online stores, craft supply stores, wholesale distribution companies, and farmers' markets will often sell dried mixed bouquets and single-variety bunches that are ready for you to display or arrange.

PROCURING FRESH FLOWERS TO DRY YOURSELF

Purchasing flowers from a farmers' market can be a great way to enjoy fresh flowers throughout each season and an ideal place to find local flowers that often dry well. Whether you purchase flowers fresh from a farmers' market, store, or wholesale distributor, try to get the freshest local organic flowers. Flowers that have been in transit or in a store for a week may already be past their prime for drying. Once you have the flowers in your possession, do not wait too long before hanging them to dry. Filler flowers that are common in most mixed fresh bouquets, such as statice, baby's breath, or eucalyptus, dry wonderfully and allow you to get more out of a fresh flower bouquet.

WHY ORGANIC FLOWERS?

Full Belly Farm has been using organic practices since the beginning and got certified early on by California Certified Organic Farmers (CCOF) because my family wanted to create an environment that would be sustainable for future generations, and provide a safe space for animals to live and humans to work. In lieu of pesticides, we use beneficial insects, crop rotation, and cover crops to help keep the soil healthy.

When you source flowers for drying, ask your supplier or farmers if they use organic practices and look for organic labeling. If you grow your own flowers, know that the benefits of forgoing pesticides will last for generations and help to provide and create an ecosystem that is inviting and safe for all.

FORAGING FOR FRESH FLOWERS TO DRY

In addition to purchasing flowers to dry, you might try foraging. I find some of my best inspiration in the parts of the landscape that are overlooked by day-to-day farming. Luckily, our farm is surrounded by natural habitat that is perfect for foraging wild grasses, rose hips, and weeds that live outside of the tidy rows and fields. Many of the drought-tolerant plants that thrive in California also dry incredibly well. Bay, native sages, hollyleaf cherries, and oaks all bloom and grow around Full Belly Farm and are perfect for wreaths.

If you venture out to forage, be mindful that not all plants should be collected. Harvesting invasive weeds that easily shed their seeds, for example, may result in propagating more of the weeds in undisturbed locations. It's also important to practice good land stewardship when harvesting from trees or shrubs that will take a while to grow back. Try to prune thoughtfully and let plants get established before foraging from them. If you can, plant perennials, wild grasses, and native hedgerows near your home. Not only will this encourage habitat for animals and beneficial insects, but you will be able to source from your own garden throughout the entire year for decades to come.

GROWING YOUR OWN FLOWERS

Of course, the best way to get the exact flowers you want for drying is to grow them yourself! That doesn't mean you need to propagate fifteen acres of flowers. Just as my mother started out on a small plot of land behind her house, you can use what you have available. That may be a small row or two, a garden bed, a community garden, or even a windowsill. Small spaces can be perfect for growing compact flowers that can be harvested for drying. There are many things you should do before planting seeds: research your growing zone, test your soil, make a list of your favorite dried flowers, and cross-reference that with your growing zone, for example. There is nothing quite like the excitement of growing something from seed, tending to it as it flourishes, and then harvesting it with your own two hands.

THE FARMER
AND THE FLORIST

When I began my floral business, I was adamant that I didn't want to be a farmer. Farming seemed difficult and filled with tasks that came across as unromantic and practical—attributes that I rejected from a young age. I decided I much preferred to be wooed by nature. I wanted to have a passionate romance with each flower, and I spent the first years at the farm after college hiding from the parts of our business that I viewed as overly "ordinary." My mother and I have always joked that she is the *farmer* and I am the *florist*. The farmer is sensible and realistic, while the florist is artistic and idealistic. A seasoned farmer knows the limits of the soil, seasons, and the stewards themselves. The farmer's way to undertake growing flowers may be to develop a sales and marketing strategy, measure your space and create a field map, and check the variety and growing instructions before purchasing seeds. These simple steps and important lessons took years for my fickle heart to lean into. The florist in me wanted desperately to buy seeds with reckless abandon, plant them with hope and optimism (and very little planning), and harvest armfuls of whatever managed to live. As I got older, my marvel at each step of a flower's life began to stir something deep inside of me. My romance with the flowers seemed trivial without touching the seed, tending to the plant, or learning the history behind the blossom. I started to not only embrace the technical aspects of cultivation but also learn to enjoy the difficult but rewarding processes that I had shied away from in the beginning. In fact, I now identify myself as a farmer, and since I have returned to the farm, my mother has had the time to slow down and be romanced by the flowers and has nurtured the florist within herself once again.

There is a juxtaposition between these two mindsets, "the farmer" and "the florist," and while it's important to see the difference between these ways of undertaking tasks, it's even more important to see how they can unite to create a complete way of living. As you move along your floral journey, you may be drawn more to one way of working with flowers than the other. There is no correct way—do what feels right for you and learn to explore the spectrum of possible ways to work with flowers.

HARVESTING

When we harvest flowers for drying, we look for the peak of the flower's or grain's bloom, right before they begin their descent into seed. Each flower has a unique point at which it will hold its color best and shed its seeds or petals the least. In the second section of the book, I give pointers on how to tell when each flower is ready to harvest, but some flowers may take a bit of experimentation on your end.

Unlike their fresh counterparts, flowers for drying do not need to be harvested early in the morning. In fact, it's nice to give them a bit of time for any dew to evaporate so the bundles dry faster. We bunch all our flowers for drying out in the field, using rubber bands and stripping any leaves about 1 inch below the neck of the flower as we harvest, before laying them in the furrow to collect on our walk back. The bunches change depending on the shape and size of the flower, but we don't put

more than twenty stems into each bunch, ensuring they fit easily into our hands. Once we have finished the harvest, we carry the bunches in from the fields in giant armloads or by using crates or baskets.

Depending on where you live, you may find that smaller bundles are more effective as they will take a shorter amount of time to dry. Make sure you don't overpack your bunches or they may mold or lose their color from taking too long to dry. The most important part is that the bundles get hung within a day of being harvested so they maintain their structure and form.

A NOTE ON GREEN VERSUS MATURE

Harvesting flowers, grains, and grasses at different stages of maturity can produce completely different appearances and give you more variety in your dried arrangements. Many grains and grasses are beautiful when harvested at their gray-green stage but are just as striking once fully mature. Experiment with drying grains and grasses at various stages to see how the color and textures change as they progress. Take care to let the seeds fully develop before harvesting to ensure the plants don't mold or turn brown.

We leave about 50 percent of our grains to mature fully so we have a nice variety, but you may learn you enjoy working with one over the other. With the right weather, some grains and grasses dry naturally out in the field without the need to hang-dry them. Crops such as milo, broom corn, wheat, and flax are all examples of grains we let dry naturally in the field once we have harvested from them earlier in the season.

DRYING

The technique of drying flowers may seem overly complicated, requiring a lot of materials. But the first time that I experimented with the process on my own, I was blown away with how simple it could be. I had moved away from home and longed for the smell of the Wreath Room and fresh flowers from my mom's garden. When she came to visit in the spring, she brought a bouquet of peonies from our farm, and I couldn't bring myself to throw them away after she left. So, in desperation, I poured the water out of the vase and left them to their own devices. Lo and behold, a couple weeks later, I had a beautiful bunch of dried peonies that I kept on my bedside table for years to remind me of home. I think everyone may have a similar story of this lazy-drying method, and, in all honesty, I still experiment with drying flowers this way, whether out of forgetfulness or curiosity. The process of drying flowers really isn't that much more complicated than that.

There are countless ways to preserve flowers, including dehydrators, silica gels, and pressing flowers. Hang-drying flowers seems to not only be the longest-standing practice but the simplest as well. There are only five things you need when drying your own flowers at home: a semi-dark room, warmth, flowers, something to hang your blooms from, and a bit of time.

WHAT TO DRY

While you can dry every type of flower, herb, grain, or grass, not everything dries well. At Full Belly Farm, we look for three main things when we select plants to dry: something that holds its color, does not easily shed its petals or break, and has a unique shape, texture, or color. Part Two of this book is devoted to everything we dry and provides helpful tips and techniques for knowing when to harvest and dry each plant. We have learned which plants dry well in our region through trial and error over the last forty years—a process I have lovingly come to call "dry-aling." We have also learned that many of our favorite plants, while beautiful when fresh, do not hold up after they are dried. Some preserve their color beautifully, whereas others turn brown or shed their petals. You never know what will work or what may excite you until you hang it upside down and test it yourself.

MATERIALS LIST FOR DRYING YOUR OWN FLOWERS

It doesn't take much to dry your own flowers aside from finding the right space, but having the right tools for the job can make the process that much easier. You can easily source these materials in any hardware store, and you may even have what you need hanging around at home.

CARDBOARD STORAGE BOXES

Once your flowers are dry, you may want to store them in boxes until you are ready to use them. We use cardboard boxes because they allow for some ventilation, which decreases the chance of molding. Choose cardboard boxes that have a solid top and bottom and that are wide enough for your bunches to fit comfortably without smashing against the edges.

CLIPPERS

A sharp pair of clippers that are lightweight are best for harvesting.

A DARK ROOM WITH GOOD VENTILATION

Overexposure to sunlight can cause flowers to fade in color, so a semi-dark or low-light room is best for drying flowers. Ventilation is key to drying flowers quickly and keeping them from molding. A box fan or dehumidifier can work in rooms with little airflow.

A FLOOR THAT IS EASY TO CLEAN

Dried flowers can shed petals, leaves, and seeds as they dry. It's best to use a room in your home that is easy to sweep.

RUBBER BANDS OR TWINE

We use rubber bands to secure our bunches of flowers because the stems contract as the bunches shrink during the drying process, but you could use twine instead.

WIRE

Select something strong to support multiple bunches of flowers as they hang-dry. We use 14-gauge wire and string it between beams or in between nails on a wall.

CREATING A SPACE FOR DRYING FLOWERS

The best places for drying flowers do not have to be overly complicated. To dry fully in place, plants need a space that is warm, has low humidity, and is out of direct sunlight. A dark place with low light is best because sunlight will begin to bleach the flowers over time. The drying room at Full Belly Farm does have windows, but its interior stays dim throughout the day despite the light that comes through the east- and west-facing windows. We keep the door open to encourage ventilation as plants may begin to mold without enough airflow.

Our drying room is not air controlled, or even well insulated, and the space itself is very simple. The only real infrastructure are the baling wires attached to each side of the ceiling. The wires are suspended from the exposed beams with metal eye rings and span the entire space. If you would like to replicate this in your own home, simply attach a strong wire between two hooks on a wall or use a ladder in your home or garage to suspend your flowers. I have also found that old wooden clothes drying racks, which can be moved and stored away easily, are a great way to affix drying flowers in a compact space without the need for permanent infrastructure.

One reason that we suspend all of our flowers from wires and hang-dry them is so that the stems and flowers stay rigid and erect in the entire dehydrating process. To hang-dry your tied bundles of fresh flowers, attach them to the suspended wires by adding another rubber band to the bunch and looping it over the wire and back on to the bunch itself so that it's secure. If you use wire, ribbon, or twine in lieu of rubber bands around the base of your stems, you may notice these ties may begin to fall off the bunch as the plants begin to shrink as they dry. The elasticity of a rubber band is useful in the way that it clings to the bundle as the stems shrink. They do break over time, though, so make sure to secure the bunches with wire or twine once they are completely dried if you plan on hanging them in place as decoration.

A NOTE ABOUT THE DRYING PROCESS

❁

Falling in love can be messy and so is the process of drying your own flowers. Seeds and petals will fall, but the sweet aroma that rises up as you sweep will remind you that dried flowers, while not a clean and tidy relationship, can provide a long love affair with blooms that would otherwise be fleeting.

Dill flower 2022

HOW TO KNOW WHEN YOUR FLOWERS ARE DONE DRYING

Each plant takes a different amount of time to dry. Some flowers, such as statice or lavender, can dry in less than a week, while marigolds or sunflowers, which have more moisture and mass, may take up to three weeks to dry. Drying times also depend on where you live. If you are in a region that tends to be humid or has cooler weather, expect the bundles to take more time to completely dry. Try adding a fan or dehumidifier near your flowers for better circulation, and plan to dry your flowers in the hottest months of the season.

The parts of the plant that retain the most moisture are the heads and necks of the flowers and the inside of the stems. To see if your flowers are completely dry, check the center-most flower in the bunch by squeezing it gently. The flower and stem should feel brittle and have very little give when you squeeze them. If you aren't sure, it's best to wait a couple more days.

STORING

After our flowers are completely dry, we take them off the wires and place them gently in large cardboard boxes until we are ready to design with them. Storing the flowers frees up our drying space, and it keeps the flowers from fading and getting too brittle. We have found cardboard boxes work best because they allow the flowers to breathe, which helps combat mold. As you place flowers in the box, make sure they are not packed tightly and alternate the direction of the bunches to allow for more cushion on the stems and flowers. Label each box with the name of the flower, the color, and the date. Store flowers out of reach of rodents and in a well-ventilated area. We store our flowers in two shipping containers, but a garage or attic would also work well.

You do not need to store your flowers if you plan to design with them right away or if you plan to keep them up as decoration in your home. We often have decorative dried flowers in our home for over a year before we begin to see any major fading. But the longer that flowers are exposed to harsh light, the more quickly they will lose their color. Dried flowers still need to be refreshed every so often, but they will last years if they are kept in a space away from direct sunlight.

ADDRESSING ISSUES THAT MAY ARISE

There are some questions that may arise when you begin drying your own flowers. I have compiled a list of some of the most common questions and concerns that come up almost every time I chat with folks about how we dry our flowers at Full Belly Farm. Each of these problems has a simple solution that can be applied to any size of dried flower operation, from a few bundles being dried at home to a drying barn at a farm. I hope these troubleshooting tips embolden you to try drying at home.

BRITTLE FLOWERS

Dried flowers have a reputation for being brittle and tricky to arrange. Over the years, we have found a couple of easy ways to soften them slightly before we begin our designs. You can easily rehydrate them by placing inside a cooler or refrigerator for a day. You can also rehydrate flowers overnight by placing a wet towel over the bunches you plan to use. It is much easier to arrange dried flowers if it's rainy or wet outside, which is why we do the majority of our arranging in the fall and winter, once there is enough moisture in the air.

DAMAGE FROM PESTS

One of the most common questions we get is about pests. Spiders, while not capable of any serious damage to your dried flowers, may be unwanted visitors. You can dust your flowers lightly with a feather duster if you are worried about spiders finding a home in your dried flowers. Mice, on the other hand, may be interested in your dried flowers—especially grains. Keep your boxes of flowers stored away from rodents. If you are concerned about potential damage from pests, a barn or outbuilding may be a better place to dry flowers than in your home. Or you can purchase flowers already dried to avoid the issue altogether.

LOSING COLOR

A loss in color is typically the result of overexposure to light or the flowers being harvested at the wrong stage. Keep your dried flowers away from bright windows and make sure you harvest before they go to seed. If you plan to keep the flowers up as decoration once dry, know that they will last longer than their fresh counterparts, but they will need to be substituted with new dried flowers eventually. We trade out our wreaths every couple years and replace dried arrangements as the seasons change to maintain their bright colors.

MOLDING OR BROWNING

Make sure dried flowers are stored in a well-ventilated area and are kept dry once they are stored away. An attic or garage is a great place to store cardboard boxes of dried flowers. Check all your bundles of flowers before storing them. Even one flower that is not completely dry can ruin an entire box of dried flowers. If you live where humidity is an issue or you begin to see mold or browning on your flowers, consider placing a fan or dehumidifier in the space where you dry flowers to circulate air and speed up the drying process.

SHEDDING PETALS

Some flowers just lose their petals more easily than others. But to avoid excessive shedding, harvest before they begin to shed their blooms—you want to catch the flowers before they begin to deteriorate—and hang them right after you harvest. The best way to remedy this is to experiment with the timing until you find what works for you in your space.

In the simple act of drying flowers, you have created an everlasting masterpiece. Whether or not you create the specific arrangements provided in this book or use them to inspire your own creations, the fact that you have preserved flowers, giving them a second chance at a long life, is enough. Dried flowers are a sustainable and unique way to enjoy flowers throughout the seasons, and adding them to your home will ensure that there is a bit of color and light no matter the time of year or what is going on in the world around you.

DRIED FLOWERS THROUGH-OUT THE SEASONS

2

My mom found flowers at a time when she desired her own role and relationship to farming that wasn't tied to row crops, being a farmer's wife, or raising children. She wanted to find her own path, create her own successes and failures, and prove to others and to herself that she could succeed as a farmer. In the beginning, she harvested flowers from her small garden and arranged them in mason jars late into the night after she had put her kids down to sleep. The first couple of years, she had just enough flowers to sell a few bunches at the farmers' markets and to her friends. Her enthusiasm and the amount of time she could dedicate to the flowers grew as her children grew. Each year, she learned from her past mistakes and planned for the next season. She slowly built her garden larger, rows longer, and dreams bigger.

As one of the first certified organic farmers in California, she carved out her spot in the floral market long before there was a focus on locally or sustainably grown flowers. As her part of Full Belly Farm expanded, so did her knowledge about floral design and proper harvesting techniques. Interns and farm crew helped her enlarge the budding business into multiple acres. Other women joined her team and began learning how to pull weeds, harvest flowers, and make mixed bouquets efficiently, dry excess flowers and seed, and create successful succession plans for the entire growing season.

I grew up learning the names of the flowers not only from my mother but from the flower farmers and interns at Full Belly Farm as well. They shared their snacks with me and kept an eye on me as I napped in the back of carts when my mom helped with the harvest. Now, I harvest alongside the flower team each day, stepping into the field just as my mom is preparing to step into some farmer's version of retirement. They have all been extremely patient with me, helping me learn Spanish and grow as a designer and farmer as I found my own role at Full Belly Farm.

I have learned how to harvest and prepare five hundred–plus mixed bouquets in a day. The "Flower Ladies," as they are lovingly known at the farm, compliment and critique my arrangements and remind me when certain flowers need to be harvested for drying. Their craftsmanship with wreaths and bouquets is unparalleled, swirling designs and color combinations that are works of art. Most of these women have been with us for more than fifteen years and are extremely hardworking and passionate. Just like my mom, they have all found their identity and a second home among the flowers. We all feel like we have the best job, and it's all thanks to a few seeds and a blooming vision that my mother planted all those years ago.

KEY INGREDIENTS

Every ingredient that we use in our arrangements falls into one of three categories—foliage and fillers, focal flowers, and grains and grasses—and this section highlights the important role that each plays in floral design. To begin to understand the functions, think of the categories as a major part of the body: Foliage and fillers being the heart, focal flowers serving as the head, and grains and grasses being the hands.

FOLIAGE AND FILLERS

As the beating heart of every arrangement, foliage and fillers provide fullness and form, which is the foundation of any design. The first task when building an arrangement is to decide on the overall shape. I like to find branches or foliage with interesting structures that manifest in the natural world. Starting with a crooked mossy branch or dramatic arching stem will dictate the direction of your design and be the spirit of the rest of your arrangement. Next, you'll build a nest with fillers and foliage. Not too dense, just a base to serve as a resting place for your other flowers and to support stems where you wish. Foliage and fillers are often the least showy part of an arrangement and may regularly be overlooked. But, like the heart, their purpose is not to be flashy, but to be strong, solid, and consistent.

FOCAL FLOWERS

These are quite literally known as "face flowers" in floral design and work best when grouped in sets of three or five. With their larger heads and eye-catching appearance, they keep and move the viewer's attention. While dried focal flowers may not be as large or as naturally showy as their fresh counterparts, they still hold a power in their ability to transform an arrangement from uninteresting to attention-grabbing. Face flowers are the sparkling eyes, the complex brains, and the coy smile behind each design. They captivate and enthrall. But they also have the capacity to overcomplicate and muddle an arrangement if not used wisely and with restraint.

GRAINS AND GRASSES

As the graceful hands of the body, grains and grasses can gesture grandly or signal softly. Grains provide a unique roughness and distinctive feel that fresh flowers often miss, while grasses hold a lightness and flexibility that add much-needed movement to dried floral design. Because of the rigidity of dried flowers, airy grains and grasses act as one of the most important parts of the design process, ensuring that the transition from the outside to the inside of an arrangement happens naturally and appears seamless and bringing cohesion to the final piece. Some grains, such as sorghum, are solid and dense and don't provide much movement at all. They should be used sparingly and with great consideration.

The flower guides for each season are organized in the following way to help instruct you as you begin to design with dried flowers. These guides are built to aid in seeing ingredients not only for their colors and textures but for the roles in floral design. Grouping ingredients as foliage and fillers, focal flowers, and grains and grasses will help you in knowing when to add certain ingredients in arrangements, where to place them in your final designs, and how to use them to create your desired outcome.

A NOTE ABOUT GROWING YOUR OWN FLOWERS

Farming is not a linear act. There is no cut-off day between spring and summer, so while this section is broken down seasonally, there is a natural overlap between each season. By the end of fall, the temperatures in our valley have dropped enough that it's impossible to dry things efficiently. With the frost comes the end of our drying time—which is why winter is left out of our guide to flower drying. If you plan to grow your own flowers, please note that these seasons may be different for your region.

Gardening and farming offer important lessons in patience and perseverance. You may plant and plan for a dry, warm winter only to be hit with a late frost that kills your early flowers. You may get your hopes up about harvesting beautiful amaranth only to find that the bugs or birds have gotten to them first.

Flower farming is a unique combination of overplanning, guesswork, and observations. There are countless obstacles that we cannot plan for—late frost, lack of rain, a pandemic, low germination, or pests. It's a beautiful and nerve-racking thing to put all of our trust into the earth each year; to bury it deep like a seed, hoping it finds its way to the light.

Spr

The Capay Valley hills turn a rich emerald green in the spring after winter rains. Giant valley oaks leaf out, and a thick layer of grass and small wildflowers make pockets of gold and blue on the hillside. Our fall-planted bulbs begin to emerge in early March, just as the wildflowers begin to get under way on the hills. It's still too early to begin drying flowers, so we enjoy these fresh spring blooms in their novelty, watching them exquisitely open and then wither in vases and jars on our kitchen tables.

The first flowers coincide with so much new life on the farm. Baby lambs rove green pastures—bleating to their moms as they roam and make friends. Chicks join the flocks of chickens at the farm. The first potatoes are dug, asparagus is cut, and almond blossoms pop, turning the whole valley into a bright-white, honey-scented wonderland of petals and bees. It's not until about mid-April that we have some days that feel hot enough to dry flowers successfully. It was only recently that we experimented with drying early-blooming flowers such as ranunculus and anemones—both of which we discovered, if hot enough, dry beautifully and keep their saturated colors.

In May, the drying begins in earnest. The fields of larkspur reach toward the sky, becoming a forest of perfect stems of rich blues, pinks, and purples. The flax sheds its beautiful delicate-blue petals and expands into its seed pods, and the wheat berries begin to turn from green to gray to brown just as the valley hills do the same. We begin to get our first warm weeks, allowing us time to fill, take down, and refill the wires in the awaiting Wreath Room. Spring feels like a budding opportunity to implement new ideas with enthusiasm. The possibilities seem endless, and the moments in the Wreath Room are no different. But everything in the spring is done with care and curiosity before the heat and ferocity of the summer begins.

Spring Flower Guide

FOCAL FLOWERS

1. ANEMONES

Dry once the flower has fully opened. Blues, pinks, and purples dry well and keep their punchy tones.

2. LARKSPUR

One of our favorite dried flowers, larkspur comes in a variety of colors. Harvest when the tips of the blooms are still in bud to ensure they don't lose their petals. Larkspur is a unique flower that is used occasionally as a filler flower but can be a focal flower when the colors are bright.

3. RANUNCULUS

We harvest our shortest-stemmed ranunculus for drying. We have found the larger heads and brightest colors dry best.

4. ROSES

Although we don't grow many roses, I can't help but dry a few from my mother's garden every year for flower confetti and potpourri. Dry before the flowers have fully opened to avoid shedding.

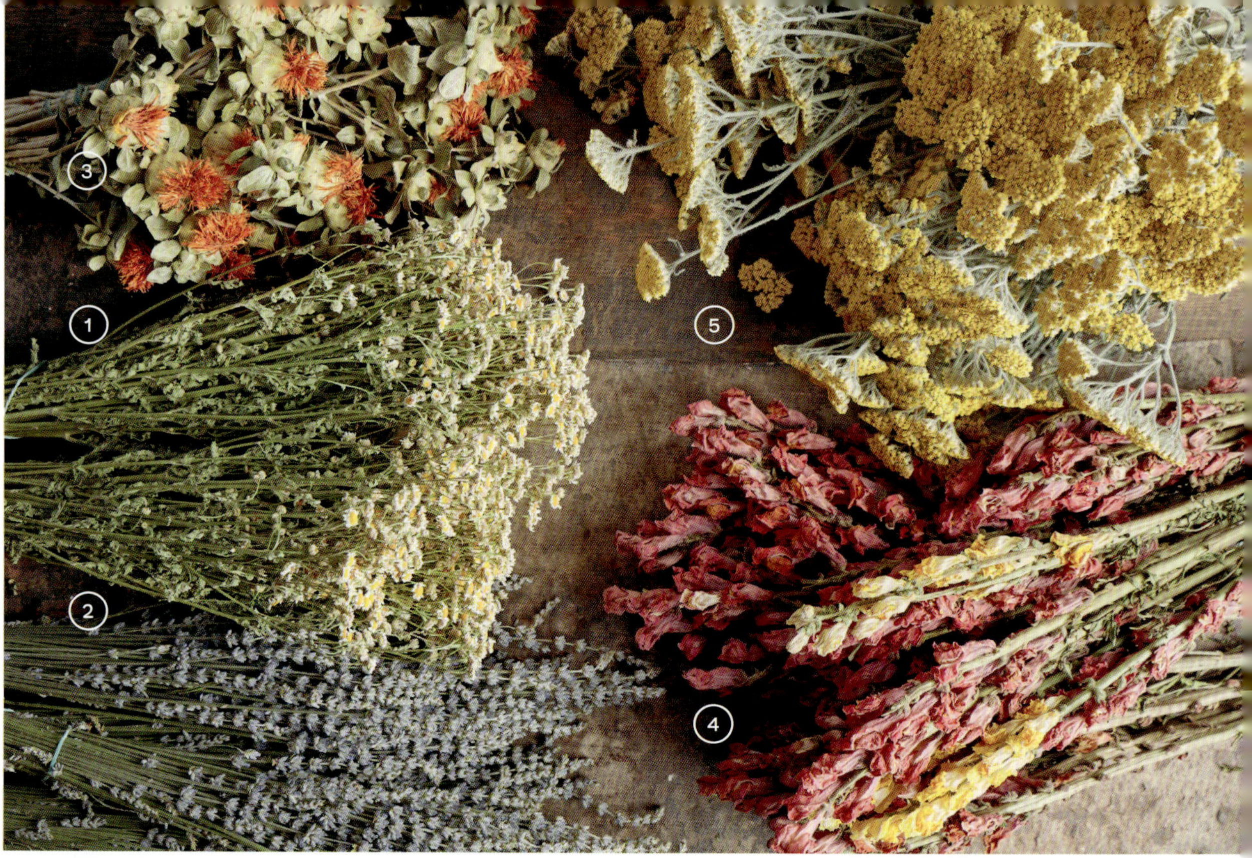
3
1
5
2
4

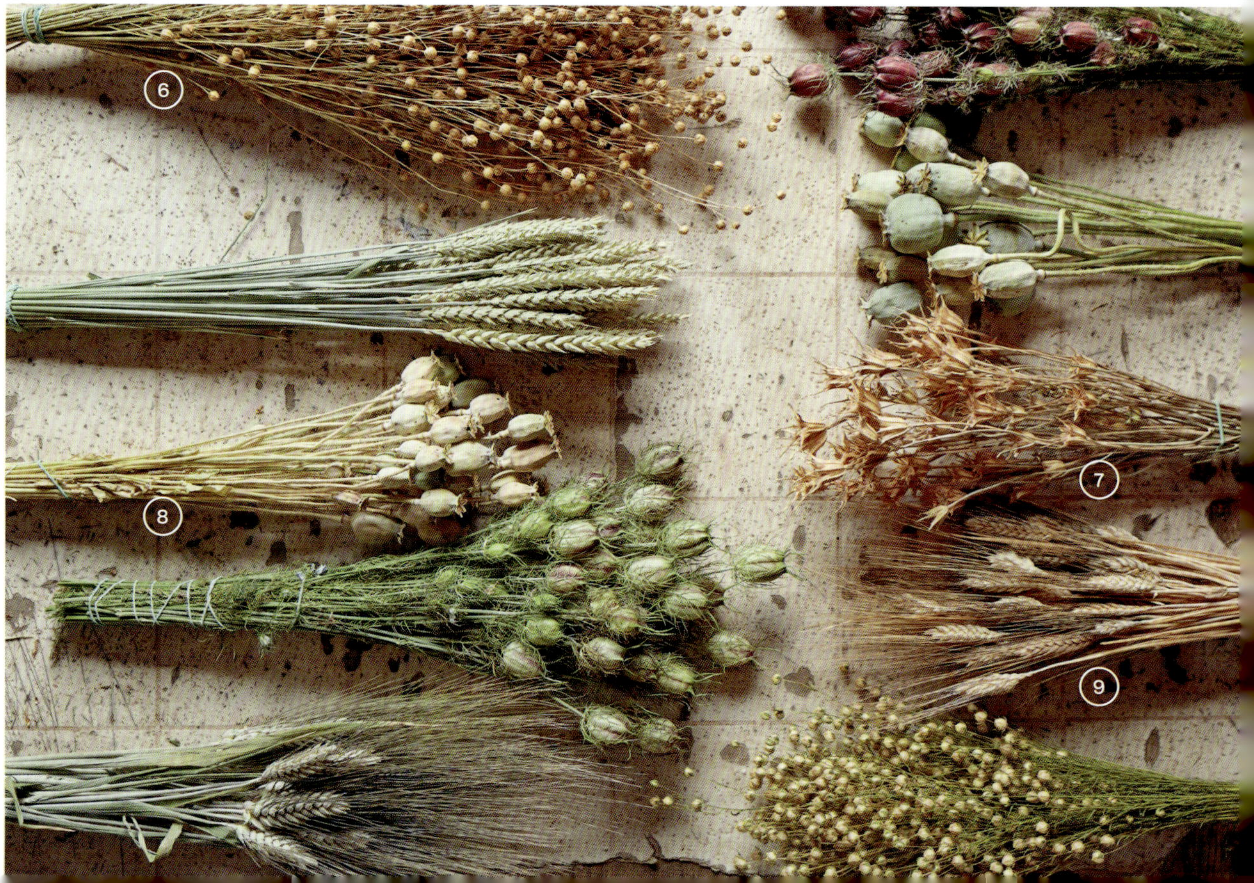
6
7
8
9

FOLIAGE AND FILLERS

1. FEVERFEW

Feverfew is such a delicate and beautiful daisy-like flower. Dry before the center begins to brown. All varieties dry beautifully but our favorite is ultra double white.

2. LAVENDER

Not all types of lavender dry well. Choose a variety for deep color as some shed their petals and fade quickly. Try English lavender Royal Velvet or Miss Katherine, or French lavender varieties Gros Bleu or Grosso for deeper blues. Harvest right before the tips of the flowers have opened.

3. SAFFLOWER

The same flower that is grown for cooking oil can also be grown as beautiful foliage for dried floral design. We harvest it at two stages: the green budding stage before color appears and once after the plant blooms.

4. SNAPDRAGON

We harvest our bright snapdragons once they have fully bloomed. The yellow and pink make beautiful additions to wreaths.

5. YARROW

This plant should be harvested once the flower has completely opened. Yellow varieties keep their color best.

GRAINS AND GRASSES

6. FLAX

We harvest our flax both green and mature. If you harvest them when they're mature, wait until the stems are ridged, they have lost all their flowers, and the pods have fully formed. Most grains, once dried, are of great interest to mice. Take great care when storing to ensure they are away from pests.

7. NIGELLA

We grow many different varieties of nigella, which comes in a variety of colors and pod shapes. Nigella 'Transformer' and love-in-a-mist are some of our favorites as they both create beautiful and unique seedpods perfect for design.

8. POPPIES

We grow poppies specifically for their pods. The flowers, while beautiful, do not dry well. Harvest when the pod begins to rattle and starts to turn from light gray to brown.

9. WHEAT

While there are many varieties of wheat that work for drying, our favorite is black awned wheat. We harvest it for its silvery foliage while it's still green and again when it has dried naturally in the field.

Sum

ur summers in the Sacramento Valley are extremely unforgiving and a true testament to the strong wills of the farmers. Temperatures creep steadily higher at the end of May, before settling somewhere between the high 90s and low 100s for the rest of the season. We all have different ways of coping with the sun. Large animals find shelter below shaded trees, the chickens bury themselves deep in the dirt, the pigs rely on giant wallows of mud, and the humans drink cool beers, munch ice-cold watermelons, and depend on a dip within the slow-moving creek to clear our heads.

While the temperatures make us dizzy and the hills brown and our creek begins to dry, it is also perfect weather for growing sweet juicy melons and giant heirloom tomatoes, and for drying flowers rapidly and efficiently. As slow as the heat makes us, it's a fast-paced time at the farm. In the beginning of June, we start work before the sun rises. We harvest flowers and vegetables promptly at 6:30 a.m. to escape the heat and give the flowers their best chance at survival. Truckloads of flowers get driven into the shed and quickly carted off to the flower cooler.

We dry most of our flowers in June through September, when the amaranth begins to grow tall and zinnias flower without inhibition. Things dry within days during these hottest months. This means that the limiting factor isn't space on the wires in the Wreath Room, it's finding enough time in the day when we aren't all rushing to complete our fresh flower orders or trying to get out of the fields before the sun hits the middle of the sky.

Escaping the day-to-day chaos of the farm in full swing to hang flowers in the Wreath Room feels a bit like entering an altered reality. Tired eyes and constricted pupils from the bright sun take a bit of time to adjust to the darkness of the room. Even with its uninsulated walls, the departure from the sun's rays is a much-needed reprieve. The colors and vibrancy of the flowers dried in the summer are unmatched. There are flowers hung on every available wire and nail to make room for all of the bounty. But with the heat, in no time at all, they will be removed and stored, and new flowers will be rehung in their place.

Summer Flower Guide

FOCAL FLOWERS

1. CRASPEDIA

Harvest when the ball begins to turn completely yellow but before it begins to shed pollen.

2. MARIGOLDS

The biggest marigolds of the season will dry the most beautifully. We dry both the yellow and orange florets. If you plan to make a garland with them, make it while the flowers are still fresh (see page 135).

3. RUDBECKIA

Dried black-eyed Susans and yellow rudbeckia tend to fade more quickly than other flowers. Make sure you store them away from sunlight until you are ready to use them. Double-flowered varieties dry best.

4. STRAWFLOWER

Harvest as the stem begins to harden and the flowers are just beginning to show their center—the flowers tend to open more as they dry.

5. SUNFLOWERS

Not all types of sunflower dry the same. Some varieties shrink dramatically or the petals fall off. We grow a variety called Vincent's Choice and dry them before the flower has begun to shed its petals. Dry in the hottest months of the year.

6. ZINNIAS

Large double-flowered and cactus varieties dry the most beautifully. Cut when the neck becomes rigid and the flowers are in full bloom.

FOLIAGE
AND FILLERS

1. AMMOBIUM

Harvest once the center flowers have begun to yellow, most of the flowers have begun to open, and the stems stay ridged when wiggled. Cut the whole plant at the same time for a successful second flush.

2. BABY'S BREATH

Harvest once the flowers have fully opened or dry naturally in the field for a more sun-bleached appearance.

3. DILL

Harvest as the plants begin to harden and turn white, or wait for it to naturally dry and brown in the field for a more mature and bleached look.

4. STATICE

Harvest once the flowers have fully opened and before any browning occurs. Statice can be brittle when dried, so it's best when used fresh in dried designs or rehydrated before use (see page 42).

GRAINS
AND GRASSES

5. MILLET

We grow many different varieties of millet and dry them at various stages for green and mature looks. Some of our favorite varieties are foxtail, limelight, and the wispy juosves. Birds and mice also love these tasty grains. Be prepared to share with your feathered friends and plant more than you think you'll use.

6. TASSEL AMARANTH

We grow more than five different varieties of tassel amaranth. Harvest once the spike has begun to form but before it starts to go to seed. We use the central stem as well as the leaves in dried arrangements and wreaths. If you harvest early in the season, you can expect to get another flush from your plants.

Fa

I t's the cool nights that first begin to hint at the arrival of fall. Every year, it appears later than we remember, with early-October days generally still so scorching that they leave us breathless. Busy tasks continue to fill each waking moment, not much different from the frenzied midsummer energy. For us, fall isn't marked by golden leaves on maple trees, lattes sprinkled with nutmeg, or breezy commutes that call for tights and scarves. Instead, it's evident in the first pomegranates ripening on the gnarled trees that line the roads around the farm and in the persimmons turning from green to rich orange. We glimpse it in the first dusty-pink morning, when we can see our breath as we harvest and, most excitingly, when the first rain of the season arrives and splits the last of the tomatoes in the field and soaks the parched soil. It allows all farmers to take a collective breath and stop for a moment in thanks to look up to the sky from our earthly tasks.

It's my favorite season on the farm. As we await the first frost that we know will brown the flowers in the field and bring an end to the last zinnias and marigolds, we also see the last hurrah of summer. The flowers seem to shove all of their final energy into producing the largest and most colorful blooms of the year. The cosmos put on their best show, and the celosia gleams bright against the weedy fields. The farmers also take part in this celebration with harvest festivals and dinners that last late into the night. And finally, when we are least expecting it but fervently willing for it to come, the first frost hits, and with the sudden browning of the flowers, the fresh-flower season draws to a close in just one day and our dried-flower season begins. The storage containers that hold a year's worth of preparation are pried open, the doors flung wide, and our fingers greedily reach for the flora that we so deeply missed.

Fall Flower Guide

FOCAL FLOWERS

1. CELOSIA

We grow countless varieties of celosia for drying. Some of our favorites are feathered, spike, and plume varieties. We especially enjoy varieties flamingo feather and terracotta. For the brightest colors, harvest before the heads begin to shed their seeds.

2. COCKSCOMB

Red, burgundy, green, and dark-orange cockscomb dry beautifully, while some pinks, yellows, and oranges turn brown. Harvest before the heads begin to shed their seeds.

3. GLOBE AMARANTH

We harvest when stems are fully stiff but before the flowers begin to shed their petals. Red blooms seem to shed more than other colors.

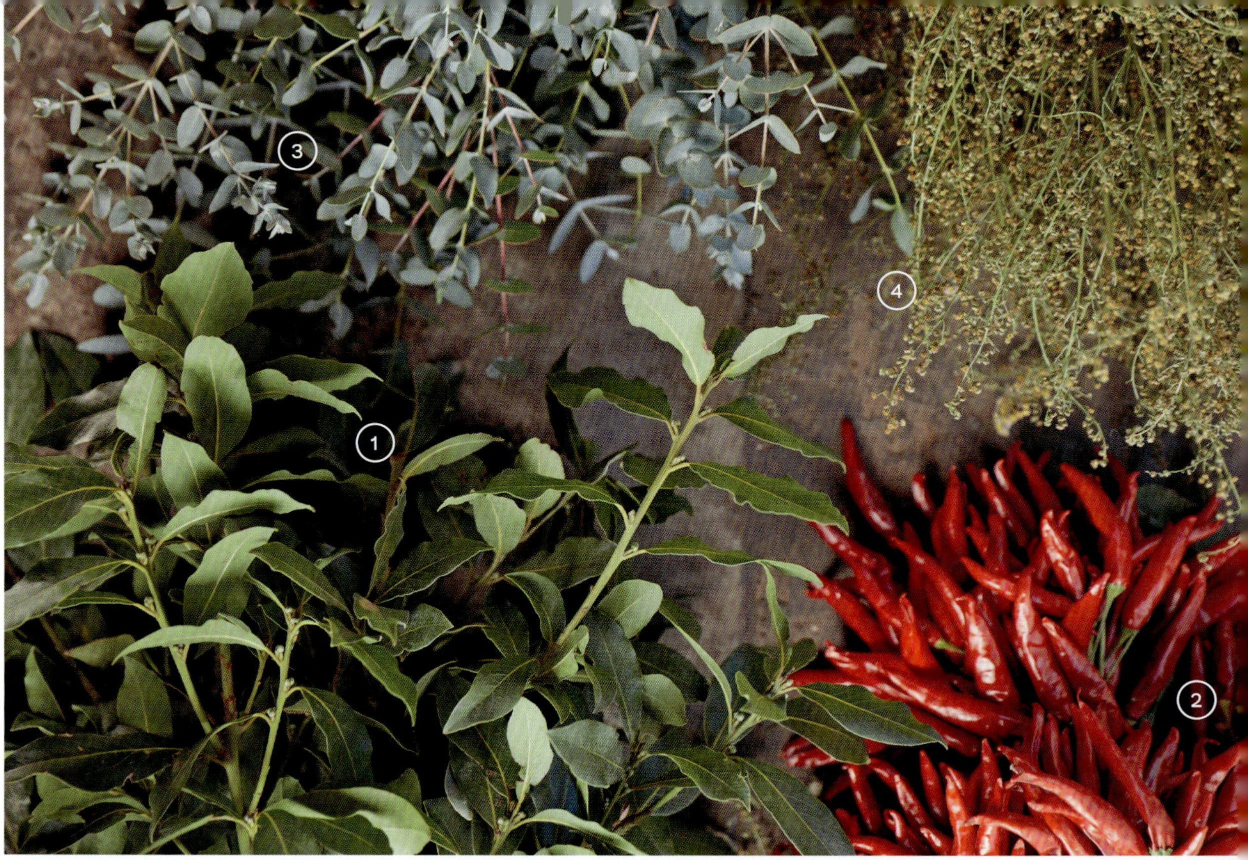
3
4
1
2

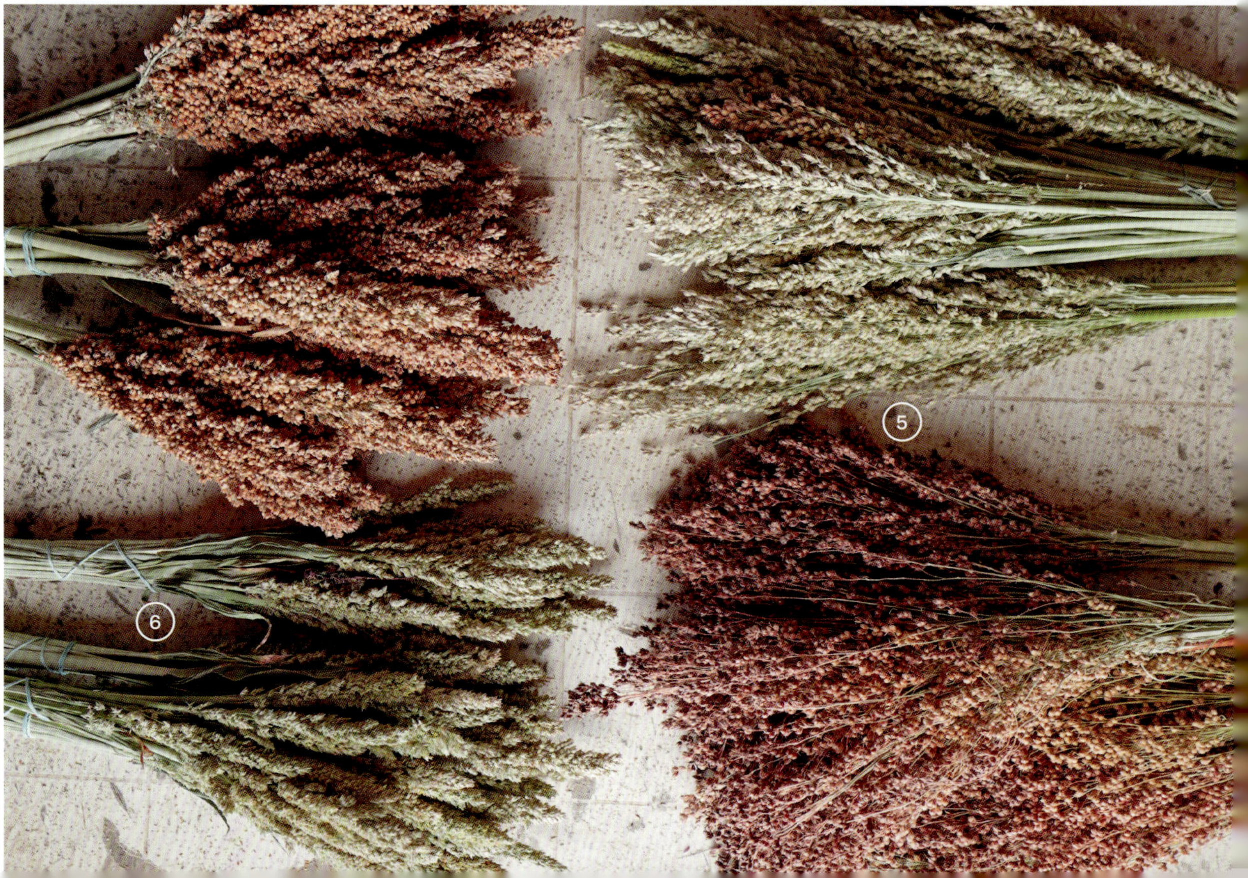
5
6

FOLIAGE AND FILLERS

1. BAY

Bay is one of the best types of green foliage to use in wreaths as it's edible and keeps its luscious deep-green color once dried. Use fresh in arrangements and let it dry naturally in place.

2. CAYENNE CHILES

Harvest once all the chiles on all of the stems have ripened to a deep red. Use fresh or partially dried for the best result. The chiles are spicy, so use caution when harvesting and designing with these.

3. EUCALYPTUS

Use fresh in wreaths and arrangements, as it's easier to work with and will dry beautifully in place. Harvest once the greenery has matured and the tips of the leaves are no longer soft.

4. SWEET ANNIE

Begin harvesting when yellow balls of pollen appear on the ends of the foliage; this may require vigilant checking to make sure you don't harvest too early. Many people are allergic to this flower, so plan and plant accordingly.

GRAINS AND GRASSES

5. BROOM CORN

If harvested early, broom corn will produce a second flush of stalks. Harvest green or mature for two very different looks.

6. MILO

Harvest in its early green stage and your second flush will give you smaller, more manageable stalks to harvest when they are fully mature.

DRIED FLOWER ARRANGE-MENTS

3

Flowers have always been a constant in my life. They act as a gentle reminder to observe my surroundings, take note of the seasons, and embrace the beauty in the smallest things. I hadn't even realized how many ways I incorporate them into my day-to-day life until recently, when I stopped to reflect. I routinely collect wildflowers and leaves and press them between the pages of my favorite books, and although I often quickly forget them, I know that when I return to those familiar pages, they will surprise and welcome me. Strands of dried rosebud, marigold, and strawflower hang from various corners of my home, and bowls full of my favorite dried blooms are nestled atop tables and bookcases. Most of my floral decorations are extremely simple. These modest designs are a reminder that things don't need to be complicated in order to be beautiful. Yet there are moments in life when grander, more elaborate designs feel better suited. The dried floral wreaths that we make in the fall celebrate the rainbow of colors that we have dried and demonstrate the intricacies that are possible with enough flowers, practice, and time. I have also made dried bridal bouquets for weddings, dried floral centerpieces for celebrated dinner parties, and a myriad of different large-scale designs with the flowers we grow.

The best part about dried flowers is that you don't need many to enjoy them immensely. Even if this book inspires you to dry only one bunch of flowers, that one bunch will go a long way. The fact that you can keep working with your bundles, pick them up and move them, cut them down and re-create with them, makes dried flowers not only the most sustainable option for decoration but the most cost-effective as well.

Now that you have learned to dry your own flowers at home, it is time to practice living, arranging, and celebrating with dried flowers. This section features three methods of designing with dried flowers: Small Moments, Wreaths, and Celebrations. Small Moments, about celebrating and commemorating the commonplace, presents simple ways to bring dried flowers into your home to make each day more colorful. Wreaths fits naturally into the middle as they are at the heart

83

of what we create with dried flowers at Full Belly Farm. You will learn techniques for creating full wreaths in an assortment of styles and in a variety of colorways. And finally, Celebrations dives into creating large-scale designs for all manner of celebrations. This last section tests the limit of what can be created with enough time, inspiration, and material. Each how-to is meant to be a jumping-off point to express yourself as an artist and designer, to instill inspiration and confidence, to stir and satisfy the voice inside of you that whispers "Create!"

IMAGINING A HOME FOR YOUR DESIGN

Each time that I create with dried flowers, I try to envision the design in my head before I begin, and imagine where its final resting place might be. Will its home be a bright-red door? A cozy mantel above a fireplace? Or a table amid a garden? Do I want my creation to smell aromatic and strong? Or the flowers to be long-lasting, able to fare well in weather and with natural sun-bleaching? This process helps me decide on the color scheme and dictates the flowers I use. Each flower has a unique set of characteristics.

Designing is not just about creating a pattern but about creating for a specific place. Some designs are best kept simple, allowing the color palette and materials to speak for themselves, while others benefit from added intricacies that make specific flowers dance and become the stars. Whatever you choose, I invite you to close your eyes before you begin and imagine your arrangement's final home. This practice allows you a moment of stillness and gives direction to your final creation.

MATERIALS

Your materials, including tools and vessels, will be the building blocks for every design, large or small. The materials you use in floral design are also dependent on your vision and the space you are starting with. It's important, for example, to choose a vase or vessel that fits the purpose of your design.

Before I begin larger installations or designs, I find it helpful to briefly sketch my idea on a piece of paper. From this sketch I create a list of the materials I may need along with the amount of product I think will be necessary to complete the design. That way, I have the list and the sketch as reference as I begin my creation, and others that are assisting me can visualize the design as well.

When I pull ingredients for designs, I often overindulge and pick out more flowers than I know I need. That way, I have added choice and flexibility when I am creating. In many of the projects in this section, you will notice I pulled more dried flowers than I ended up using and chose to exclude a few types of flowers in the final creations. I enjoy being able to change my mind, adapt my designs, and listen to my heart while I create!

VASES AND VESSELS

Because dried flowers do not need a water source, you have a lot more creativity and flexibility when it comes to choosing a vessel. As you discover which types of designs you like to create, you can begin to build out your vase collection. I recommend that everyone have at least three types of vases in their home.

BUD VASES These tiny vases are perfect for simple arrangements that take minimal amounts of flowers.

A LARGE CERAMIC VASE Try working with large ceramic vases to spark creativity and match the style of your home.

AN OPAQUE FOOTED COMPOTE VASE For artistic designs, I find opaque vases easier to work with as they keep any unruly, damaged, or brown stems hidden.

Your local thrift or antique store is a good place to find bud vases. A vase that has character, is homemade, or that has a backstory is a lovely way to showcase a few of your favorite dried flowers.

If you have a limited number of flowers, try using vases with smaller openings. If you want to create grand arrangements that will be the focus of the dinner table or entrance table, try using a vase that is large with a wider mouth. Whatever you choose, your final arrangement of flowers should be larger than the height of the vase itself to make the flowers the stars of the show.

TOOLS FOR A FLORIST AND FARMER

No matter the composition you choose, there are a few supplies that every floral designer should have in their bag of tricks. All of these tools are easy to find at your local craft supply or hardware store.

APRON

When you are designing and setting up installations, an apron is a great way to keep zip ties, floral wire, and clippers handy and to keep your hands and other pockets free. Where else will you put all of these tools?

BALING WIRE

We use 14-gauge baling wire for our wreath and garland frames. We go through a lot of baling wire since we make all our wreath bases (see page 144), so we get our wire in bulk from our local farm supply store. Alternatively, you can use thick floral wire or old wire coat hangers to create your wreath frames.

CHICKEN WIRE

When it comes to large-scale arrangements, a florist's best friend is chicken wire. We get our chicken wire from our local tractor or orchard supply store. For larger designs such as installations, you can create armatures for your stems to rest within. You can also use crumpled pieces of chicken wire within vases in lieu of a floral frog or floral foam.

CLIPPERS

We have many different types of clippers for different stem thicknesses and design purposes. While foraging, we use loppers or bypass pruners. While harvesting flowers in the fields, we have found Felco needle-nose pruning shears to work best. Dried flower stems can be harder to cut than fresh flower stems, so keep clippers sharp. I also recommend keeping a pair with you at all times—for those unforeseen foraging or design opportunities.

FISHING LINE

We often use fishing line while stringing flowers for garlands or flower strands. Fishing line is also useful for hanging installations where you want the mechanics to remain hidden or unseen.

FLORAL FROGS

These are wonderful for keeping dried flower stems in place. Dried flowers are typically lightweight and can be tricky to keep from moving or breaking. Flower frogs, with their sharp points, cage-like appearance, and heavy weight, serve as a sturdy base and ensure that flowers stay put. The frogs are traditionally placed on the bottom of vases to keep stems in place, but I also like to arrange with them on their own (see page 107). If you use a frog within a shorter compote vase, you may need to use putty to keep the floral frog in place; larger containers will not need the extra support.

FLORAL TAPE

Green Oasis–brand floral tape is great for wrapping stems of brittle dried flowers and for creating crowns, boutonnieres, or wrapping bouquets.

FLORAL WIRE

We use floral wire for everything from flower crowns to wreath making. Our preferred gauge is 26, but 24 or 28 will do in a pinch. We have found this gauge works best in terms of flexibility and strength. It's easiest to use floral wire that is on a paddle for wreaths.

RIBBON

While ribbon isn't a necessary design tool for arranging, hand-dyed silk or chiffon ribbons add a beautiful touch to dried flower bouquets, wreaths, or crowns.

RUBBER BANDS

We use elastic rubber bands to secure all of our flowers when we dry. There are many different styles and thicknesses that work. To hang certain grains or heavy flowers, use a thicker band that can hold more weight and can be reused.

STAPLE GUN

A staple or nail gun is useful for creating lightweight dried flower installations or attaching chicken wire to a wooden structure.

WATER TUBES

For large installations or arrangements where both fresh and dried flowers will be incorporated, water tubes are useful for keeping fresh flower stems in water. There are various sizes and styles to fit your needs.

WIRE CUTTERS

Although clippers could work to cut most small floral wire, wire cutters are best for cutting heavy wires, ensuring that your clippers and loppers stay sharp.

ZIP TIES

You can never have enough of these on hand for securing large installations or attaching chicken wire to frames and arches.

HOW TO BREAK DOWN LARGER FLOWERS

It might be difficult to imagine how to use large flowers or grains in smaller wreaths or arrangements. While tassel amaranth or giant cockscomb can be perfect for installations or larger creations, it's necessary to cut them down to more convenient fragments for smaller compositions. We end up cutting or breaking down many of the flowers we grow into smaller sections so we can use more of the plant and get a piece that is easily manageable. To break down flowers, start at the top of the plant and, using clippers, cut each section into 4- or 5-inch pieces, then discard the stem and any unusable portions.

COLOR THEORY

Dried flowers have a reputation for being muted and dull, when, in reality, you can find every color in the rainbow within naturally preserved flowers. The more pressing issue, you may find, will be picking your colors wisely and thoughtfully when beginning a new creation. A quick glance at a color wheel demonstrates that there are countless color combinations—the possibilities for grouping hues and shades are endless. When choosing flowers to work with, I find it helpful to break down color theory into two distinct houses.

MONOCHROMATIC OR ANALOGOUS

For these color schemes, you will use similar shades of the same color or colors near each other on the color wheel.

COMPLEMENTARY OR TRIADIC

Select colors on opposite sides of the color wheel for greatest contrast.

Both of these design choices hold power, and you may find you gravitate more heavily toward one or the other. Nature is where you can find the best inspiration and practice noticing the duality and dichotomy within color schemes. A field of grass holds a myriad of monochromatic moments that you may find peaceful and calming, whereas a wildflower meadow may be home to many complementary color palettes that make you feel joyful and energized.

FINDING INSPIRATION

To this day, my design style is continuously evolving and adapting based on my capacity and curiosity to learn, my relationship with flowers, and my relationship to myself. The best advice that I can give as you wander the winding road of creativity is to practice living seasonally. This simple advice cannot be understated. If I feel stuck as an artist or human, unclear of my direction forward or imperfect in my design, I take a stroll around our farm or my home and notice the inchworm in my garden or the fruit ripening on the vine, and watch the heron stroll forward in rippling water.

To live seasonally is to recognize that everything is fleeting and also that everything will return once more. The sweet deep-red strawberry, the first cheerful yellow daffodil—each new taste, sight, and smell is my inspiration renewed. As an artist, I find this is helpful in staying true to my voice and following my own path. As a human, seasonality is a sweet reminder that everything has its natural order and flow, and practicing adaptability and flexibility is key to success. If you are feeling stuck, go for a walk, find a farmers' market, and ask the first farmer you see what's new in their fields this week. Eat their offering slowly, ravenously, and with great curiosity, and then approach your floral design work in the same way.

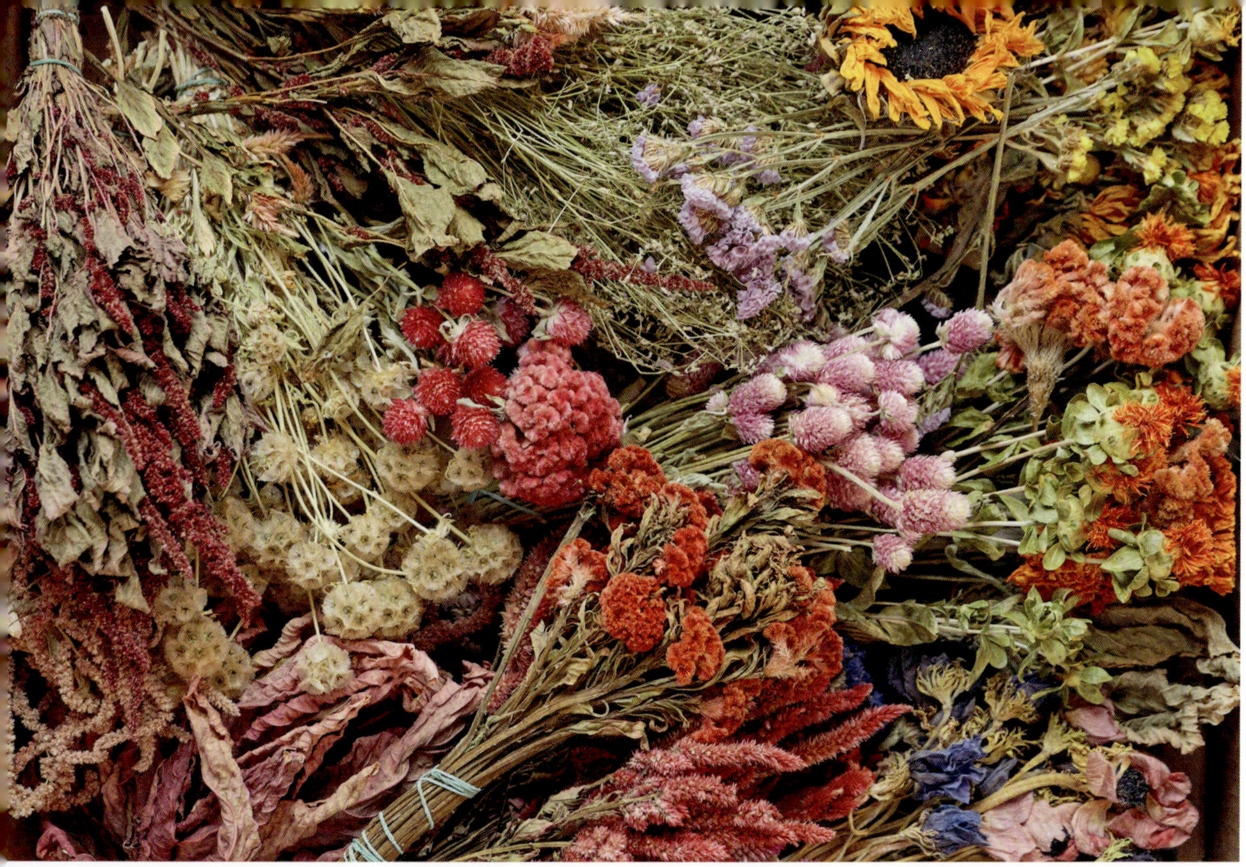

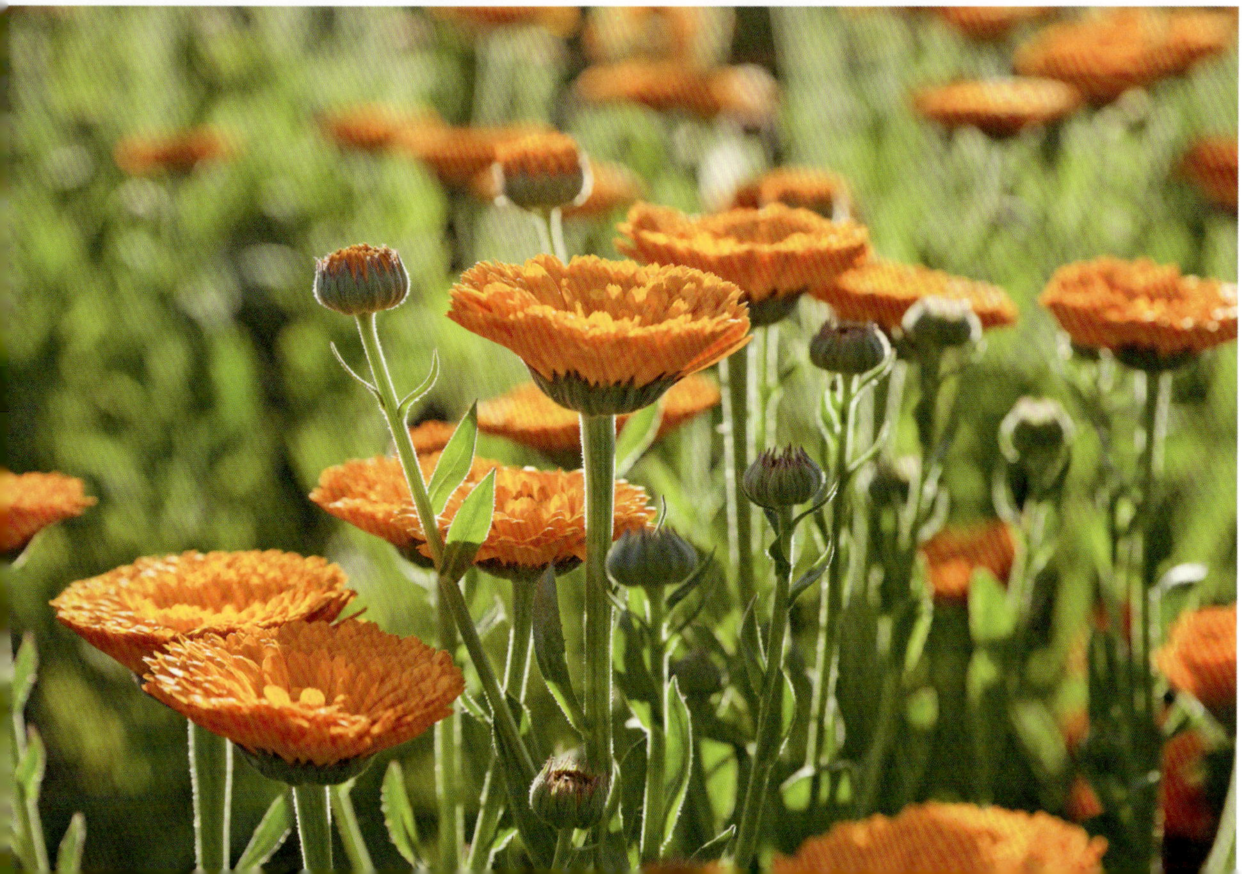

Small Moments

EMBRACING THE SMALL MOMENTS

Farming is all about celebrating the small moments. The first strawberry of the season is cause for celebration, as is the first sign of life on fields sown with care. There are many monumental moments in our lives that call for grand festivities, but many tiny moments should be commemorated as well. Dried flowers in their simplest form do not need much fussing to make them more wonderful. Merely bringing dried flowers into your home is a way to delight in something beautiful and lasting. The colorful blooms provide a burst of joy in the fall and winter when it's dark or stormy outside.

Making something small but mighty can be just as meaningful as producing a larger, more majestic piece. When I teach wreath classes in the fall, my favorite moment is when someone who came into the class claiming to be "terrible at art" leaves with something they are immensely proud of. The pride that can be instilled by just *making* and *making time* cannot be overstated. Perhaps a bedroom vanity may need decorating, or a gift could be enhanced with a small bouquet tied to the top, or a flower crown would help celebrate a Tuesday morning. This section is about those small moments that may pass us by if we don't take a moment to pause, reflect, and create something meaningful.

Bud Vases with Strawflower and Poppy Pods

MATERIALS

5 to 7 bud vases of various heights and shapes

3 to 5 filler and foliage per vase, such as dried snapdragons and feverfew

3 to 5 focal flowers per vase, such as strawflower, craspedia, and larkspur

3 to 5 grains and grasses per vase, such as poppy pods and pennycress

PERFECTION CAN BE FOUND in the simplicity of small, dried arrangements. All you need are a few flowers that will stand alone. One of my favorite ways to show off dried flowers in my home is with a small set of ceramic bud vases. Over the years, I have gathered hundreds of tiny bud vases that are meaningful and call for the simplest of flower designs. When I have dried flowers left over from another project or a few that felt too special to make their way into a larger wreath, I often sprinkle them around in small bottles or vases in my bathroom, on an entry table, or on a bedside table. Before choosing your bud vases, decide where you want to place the arrangements. That will help you decide on a height or color scheme. For these small arrangements, find flowers that have natural flow to their stems and movement in their blooms, such as a crooked poppy pod stem that branches out or a strawflower that beckons you with its bright-yellow face. This will help the arrangements feel multidimensional and draw attention to the arrangement, even if it's only a few flowers in each vase. For these arrangements, I wanted a warm wildflower look, so I chose poppy pods, snapdragons, strawflower, larkspur, pennycress, and craspedia. Aim to use 9 to 15 flowers per vase.

FOR THIS DESIGN

1. Strip the leaves and any low buds off the stems to give each flower room to breathe and space to fan out.

continues

2. Cut the flowers to the desired sizes, adding a variety of heights and textures. I always recommend that the tallest flowers be cut to a height equal to one and a half times the height of the vase, so that the flowers are the focal point of your arrangement rather than the vase.

3. Place a few focal flowers near the mouth of the bud vases, and finish off by placing a few grains or grasses gently into the vases at various heights.

4. Once you have finished, sprinkle the bud vases around your home for a wildflower look, or display them together to keep it cohesive.

Petite Gift Bouquets

MATERIALS

2 filler flowers per bouquet, such as amaranth leaves, ammobium, dill, baby's breath, and statice

1 or 2 focal flowers per bouquet, such as strawflower, zinnias, black-eyed Susans, marigolds, globe amaranth, plume celosia, and larkspur

1 or 2 grains or grasses per bouquet, such as nigella, broom corn, and flax

Floral wire

Twine or ribbon

PETITE BOUQUETS ARE some of my favorite things to create with dried flowers. These small arrangements are perfect to place on top of gifts or packages, to lay on table settings at dinner, or to use as sweet moments in décor and styling. You can tie them to the tops of wine bottles to give as gifts or use as placeholders for a special event. Mini bouquets are made very similarly to the way that I construct dried flower boutonnieres for weddings and events. They make a memorable addition to a suit lapel and can act as a keepsake long after the celebration has passed. To ensure your mini bouquet looks full, use a nice mix of shapes and textures at a variety of heights.

FOR THIS DESIGN

1. Prepare all the flowers, removing any foliage below their head.

2. Place the filler flower stems by gently holding them between your pointer finger and thumb on your nondominant hand.

3. Once you have the filler flowers in place, begin adding the focal flowers to the front while nestling in any grains to the back to act as a natural fan. Your arrangement may need a bit of finessing to get it just right.

4. Still holding the bouquet in your nondominant hand, use your dominant hand to tightly wrap the necks of the flowers with floral wire right below the base of the flower heads. Cut the stems 1 to 2 inches below the wire.

5. Tie the bouquet with a piece of ribbon or twine at the bottom, covering any of the floral wire or stems you do not want to see.

Floral-Frog Arrangement

MATERIALS

*3 to 5 filler flowers
per frog, such as
amaranth leaves
and feverfew*

*3 to 10 focal flowers
per frog, such
as ranunculus and
globe amaranth*

*3 to 5 grains and
grasses per frog,
such as foraged
wild grasses*

*5 or 6 floral frogs
in various sizes
and shapes*

OVER THE YEARS, my collection of floral-related objects has grown. I love to seek out floral frogs at thrift stores and antique stores. These uniquely shaped metals objects were popular in floral design before the rise of the harmful foam that many designers use now. While floral foam can be used only once and is not biodegradable, its counterpart can be used over and over again. Floral frogs are utilized in many ways, but my favorite option is to use them to create simple dried designs that I can place around my home. For this arrangement, as a fun twist, I opted to display the antique floral frogs without a vase, letting the beauty of these antique pieces stand on their own.

FOR THIS DESIGN

1. Prepare the flowers by removing any foliage or leaves below their heads.

2. Place all the floral frogs together in a line and begin to add the foliage and fillers at the base, starting 2 to 3 inches above the frog. The base of foliage should fill about half of each floral frog, leaving room for the focal flowers to be interspersed throughout. (I used a mixture of feverfew and amaranth leaves for a neutral base.) Various heights from 1 inch to 6 inches tall will help create depth.

3. Once the filler is in place, add in the focal flowers by placing them securely into the points or cages of the frogs. Clip them at various heights, between 3 and 10 inches. (I added 3 to 15 different colors of globe amaranth and ranunculus per frog to create a sweet blush-and-rose palette.)

continues

4.   Finish off with a sprinkle of grasses to add movement into the arrangements.

5.   Place your arrangements in areas of the house where they won't be knocked over. Keep them away from too much moisture as stems may bend or get weak with additional humidity; within the frogs, the stems have less protection and containment to keep them from weakening.

Billy Button Strands

MATERIALS

20 to 100 dried or fresh Billy buttons, amount depending on desired result

Fishing line

Needle

DRIED FLOWERS WITHOUT STEMS can be just as commanding and magnificent as ones that are still intact. In the Wreath Room, glass containers display globe amaranth blooms, de-budded lavender, and poppy seeds that shook out of their pods as they dried and look like gray jewels. In my home, I use bowls of my favorite shades of strawflower heads that broke off while drying as ornamental pieces. Each year, I decorate my Christmas tree with flower-petal-filled ornaments, packed to the brim by my nieces and nephews in the Wreath Room while I create wreaths. One of my favorite activities to do with budding florists who may want to partake in a simple floral task is to make garlands of flowers that can be hung around the home. A simple way to do this is by stringing flowers upon fishing line.

Billy buttons, or craspedia as they are also known, offer an ideal jumping-off point for home décor. Their spherical shape and cheerful yellow color make them modern and perfect for small, simple designs. When we have extra Billy buttons or when stems have broken during the drying process, we create strands with just a few simple steps.

FOR THIS DESIGN

1. Cut the stems off the Billy buttons and place them in front of you in a bowl or on the table.

2. Cut the fishing line to your desired lengths; vary the lengths depending on the look you're going for and the amount of Billy buttons you have. We made ours between 2 and 5 feet in length. Then string the first piece of line onto an

continues

embroidery needle. Tie a knot at the bottom of the fishing line so the flower heads do not fall off the end.

3.   One by one, string the Billy buttons on the strands by pushing the needle into the very center of the flower and back out the other side. Create one continuous strand or leave space in between each ball by tying a simple knot in the fishing line below each of the balls.

4.   Finish off by tying a loop in the top of the fishing line and hang them from a push pin or nail in a wall, existing beam, or ceiling. You can also loop them around a banister or create a banner over a mantel. These strands also make a perfect addition to any bedroom or Christmas tree. The bright color and modern-looking flowers will last for years.

Floral Garlic Braid

MATERIALS

*50 stems of flowers,
such as statice,
ammobium, nigella,
and flax*

*12 or more garlic
heads with about
1-foot-long, semi-
green stems*

Twine or ribbon

IN THE SPRING, the smell of garlic hangs heavy in the air at the farm. The fresh green garlic gets harvested and sorted and is brought into our packing shed to get washed of any mud. Every afternoon, the packing shed smells as if the finest French cuisine is being served. My mouth waters as I arrange mixed bouquets a couple yards away in our floral room. We have learned to appreciate garlic at every stage of its life at Full Belly Farm. Once the green tips of the shoots begin to turn slightly brown, we know that the heads are starting to form underground, getting larger and mature enough to harvest and dry for the summer. But before they dry completely, when the stems are still green, one of our favorite traditions is to make garlic braids with dried flowers. There is a reason this age-old tradition of twisting strands of garlic into art persists.

If you cultivate your own garlic for this arrangement, harvest it when there are three or four green leaves still visible. Clean off any dirt and broken leaves and trim the roots low. If you don't grow your own garlic, garlic bulbs with the greens still attached are not commonly found in grocery stores, so look for them at a farmers' market. You may need to partially dry them for a few days after you get them home so they are the right texture for braiding. To ensure that the flowers are malleable enough for braiding, use ones that have been rehydrated (see page 42) or are not yet fully dry and will dry in place.

There are a few talented individuals at Full Belly Farm who create masterful garlic braids every summer. Their technique and skill make the task look simple, but just like any craft, practice and patience is required. For this project I asked Catalina, one of our talented farmers, to step in to demonstrate how to create these beautiful designs.

continues

FOR THIS DESIGN

1. Prepare the flowers and garlic by removing any foliage below their heads. Keep flower stems longer than 5 inches and the stems of the garlic as long as possible.

2. Begin the braid by crossing two garlic heads at the neck to form an X. Place a third garlic head on top of the X and tightly loop its stem underneath and over the top of the X so that it comes all the way around to face downwards again and holds the first two crossed heads in place.

3. Add in your first flowers following the same crossed X shape so that the flower stems fall roughly in line with your crossed garlic stems. Put in only one or two flowers at a time and keep their height slightly above the tops of the bulbs. For the remainder of the braiding instructions, add in flowers as desired with each new layer of garlic heads.

4. To start creating the descending braid, add in two more garlic heads, in another crossed X shape on top of your first flowers, staggering them a half inch lower than the heads underneath them.

5. Place another head in the center of the braid with its stem pointed straight down. You should now have six heads of garlic with three distinct stem clusters.

6. Take the right stem from the bottom X and wrap it tightly across the top of your braid so that it now points down and to the left. Now you should have two layers of three garlic heads each and have formed the base of your braid. You're ready to begin a standard three-strand braid.

7. Add in another head with the stem pointed straight down and lay a second head on top, with its stem pointed down and to the left and its head up and to the right. Wrap the bottommost garlic stem that is pointed down and to the left up and around so that it joins the middle cluster. Add in another garlic head to complete another row of three heads.

8. Continue to add in garlic and flowers in the same crossing and braiding pattern until you have run out of garlic. The added heads and wrapped stems don't always follow an exact pattern. Improvise where you direct your new stems as necessary to keep each of your three braiding clusters roughly uniform.

9. With twine, tightly wrap the entire braid beneath your bottommost garlic heads. Keep an extra yard of twine on each end of your knot. Continue to braid the remainder of your stems, including the extra twine along with them. This will provide extra strength when hanging your braid on a wall as the stems dry and weaken over time.

10. When you have reached the dry ends of your stems, tightly tie up the whole braid again, using the remaining lengths of twine that come out of your braid. Trim off any dry ends of your stems and add in more flowers to the lower portion of braided stem as shown, or, for a wilder look, leave the braid exposed and the stem ends intact.

117

Simple Structural Centerpiece

MATERIALS

15 to 20 stems filler flowers, such as ammobium and plume celosia

About 20 grains and grasses, such as millet and flax

About 20 focal flowers, such as ranunculus, larkspur, and marigolds

4- to 6-inch wide-mouth vase or vessel

OUR BUSY LIFESTYLES just aren't always conducive to acquiring and keeping fresh flowers in the home; this is one of the reasons that dried flowers are more practical. I enjoy creating simple designs with four or five different elements placed loosely in my favorite wide-mouth vase or vessel. I like to use vases or containers that are opaque so I don't need to worry about the stems looking messy. I sprinkle these arrangements around my home and change them out seasonally to keep the colors bright and the displays matching the natural world around me.

FOR THIS DESIGN

1. Prepare the flowers by removing any foliage or flowers below their heads.

2. Fill in the vase with filler flowers, creating a nest of foliage for the focal flowers to rest in. I added my flax and millet at this stage as well, helping to create the base with these fluffy grains. The tallest filler flowers should be a few inches taller than the vase so that they rest close to the top of the vase. I like to keep the nest of filler flowers in a loose bowl shape, with the center being slightly shorter than the two outer sides.

3. Once you have a cozy nest of filler flowers, begin adding the focal flowers. Keep in mind that focal flowers look best when added near the mouth of the vase and branching away from the vase at various heights.

continues

4. Add the focal flowers in sets of three or five to ensure that the arrangement does not have too much symmetry, which can lead to stale designs.

5. Finally, once you are satisfied with the placement of the focal flowers in both the front and back of the arrangement, fill in any gaps and add movement to the design with a few sprigs of dried grasses.

6. Display your final arrangement in your home, adding and taking away materials as you desire.

Forever Flower Crown

MATERIALS

26- or 24-guage floral wire

20 to 30 filler flowers

10 to 15 focal flowers

10 to 15 grains or grasses

Ribbon

Pink crown: baby's breath, ammobium, and ranunculus

Yellow crown: craspedia, ranunculus, and pennycress

Orange crown: marigolds, ranunculus, and poppy pods

GROWING UP WITH a 350-acre backyard had its perks. My three siblings and I were wild and free. We climbed the giant walnut trees, had endless summer fun in the creek, and built flower-field forts. One summer activity that kept me busy for hours was making flower crowns on our grand explorations. We used the crowns to dress up for a farm party or just to celebrate an ordinary day. After the celebration, hang the crown in a visible spot to remind yourself that even small achievements are worth celebrating.

Now that my nieces and nephews have taken over the role as the farm's "pioneers," it is only fitting to adorn their wild-haired heads with flower crowns as well. To see them exploring the same rows of flowers, climbing the same trees, and enjoying the free-range life that mirrors my own gives me hope that the flower-crown tradition will live on for generations to come.

Flower crowns are made in many different ways. The bases can be bought, the gauge of wire can change, and the mechanics and techniques can vary slightly. If you prefer to purchase your flower crown base premade, your local craft supply store will have material that will be suitable for you. Some people prefer to use floral tape to secure their flowers. Floral tape sticks to itself as you tighten, so make sure to pull it taut and wrap it slowly, overlapping a bit each time you go. The method following has been a simple way of using the floral wire we already have on hand for wreaths and other projects. Be sure to rehydrate your flowers (see page 42) before you try to create your crown; if you don't, the stems can be too brittle to work with and may end up breaking in the process.

continues

FOR THIS DESIGN

1. To make each base, take a 3-foot piece of floral wire and double it up, twisting it slowly around itself, creating a loop on each end.

2. Prepare the flowers by cutting the stems down to 2 to 3 inches and placing them in piles in front of you. To make the crown less bulky, I strip the flowers of any foliage or flowers below the neck of the buds.

3. Wrap the floral wire four or five times right below the loop on the left side of the crown so that it is secure. Add the first two flowers with their stems facing the right side of the crown, then wrap tightly with the attached floral wire.

4. Continue adding one or two flowers at a time right below the heads of the first flowers, wrapping the necks of the flowers as you go, with all of the stems pointing to the right.

5. Once you have finished attaching flowers to the entire base, cut the wire and wrap it securely in front of the loop on the right side of your crown. Then bend the wire to match the shape of the head that will wear the crown.

6. Finish the crown by tightly tying ribbons through the loops on either side. Tie the ribbons together to secure the crown to your head.

Blooming in Blues Flower Branch

MATERIALS

1 large branch

About 100 stems, in a mix of filler, focal flowers, and grasses, such as blue larkspur, anemones, baby's breath, and globe amaranth

Floral wire

AS MY HUSBAND AND I walked the roads around the farm on one of our date nights after work, beer in one hand, clippers in the other, I couldn't help but notice the ways the trees had turned dormant again for the winter. The anticipation for blooming branches in the spring and delectable peaches makes it almost impossible to find beauty in the crooked branches in the dead of December. But there was one branch in particular that caught my eye. As I lugged it home, my mind began to race with ideas of what it would look like hanging heavy with dried flowers. For this project, begin by searching for your own perfect branch—one that calls to you and has unique twists and turns that make you excited to create something beautiful. The size is up to you and the space you would like to fill. For this project I found one that spanned about 4 feet. This branch would be the perfect addition behind any potluck or thanksgiving dinner buffet or tucked into the corner of a bedroom or living room.

FOR THIS DESIGN

1. Place the branch in the area where it will be displayed so you can make sure it fits and see how all the elements look in the space. Remove any unwanted twigs or stems that seem too unruly.

2. Next, prepare the flowers by stripping off any foliage or flowers that you think should be removed. Make 20 to 30 small bundles of flowers, about 3 to 10 stems per bundle

continues

(see Petite Gift Bouquets on page 103). First add a few filler flowers to the bundles, one or two focal flowers within them, and then one or two grasses slightly higher to create movement.

3. Tightly wrap each bouquet with floral wire once or twice at the neck of the flowers.

4. With another small piece of floral wire, attach the bundles to various twigs on the branch. Wrap the stems of the bundles tightly to the branch to create the appearance that the branch is blooming.

5. Install the branch in its final spot. Stand back to view the branch from afar, adding or altering bundles as needed.

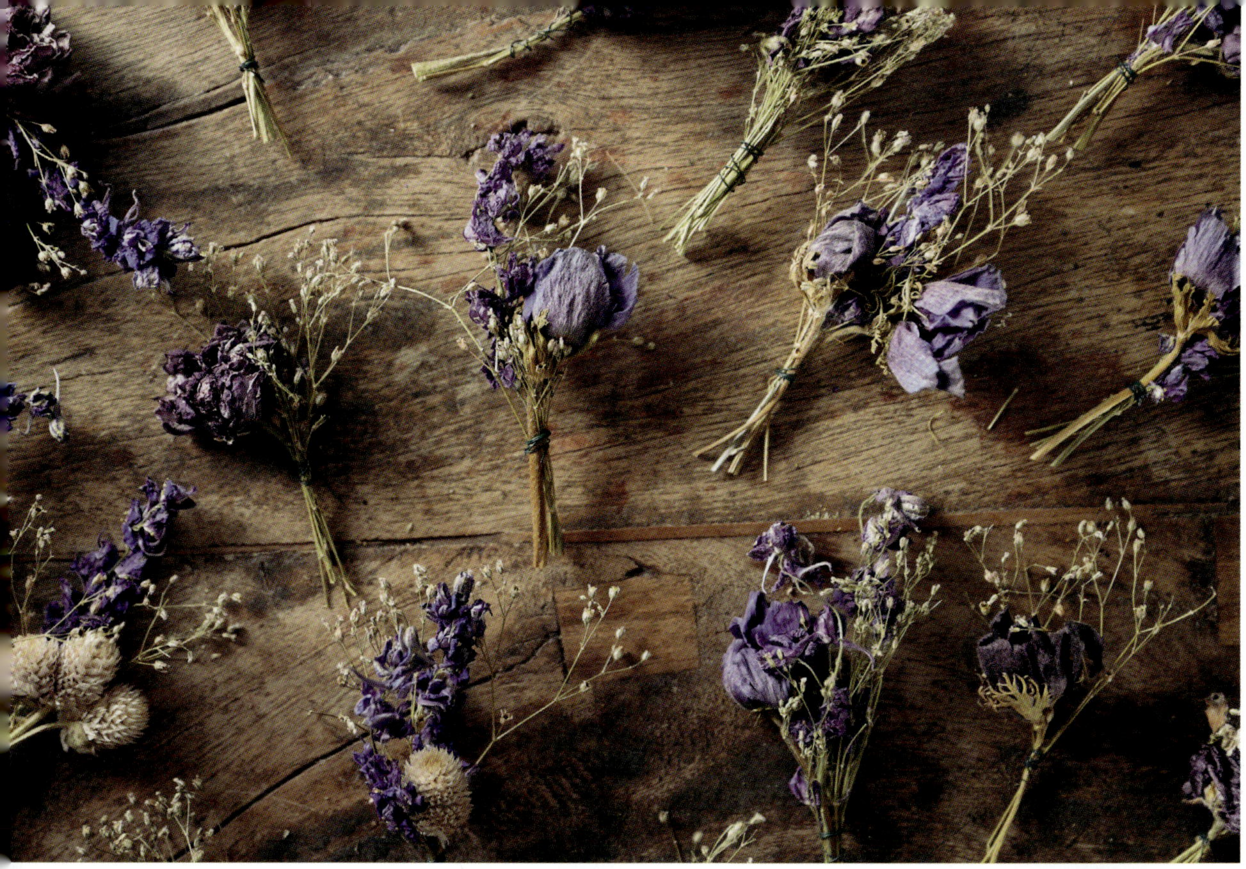

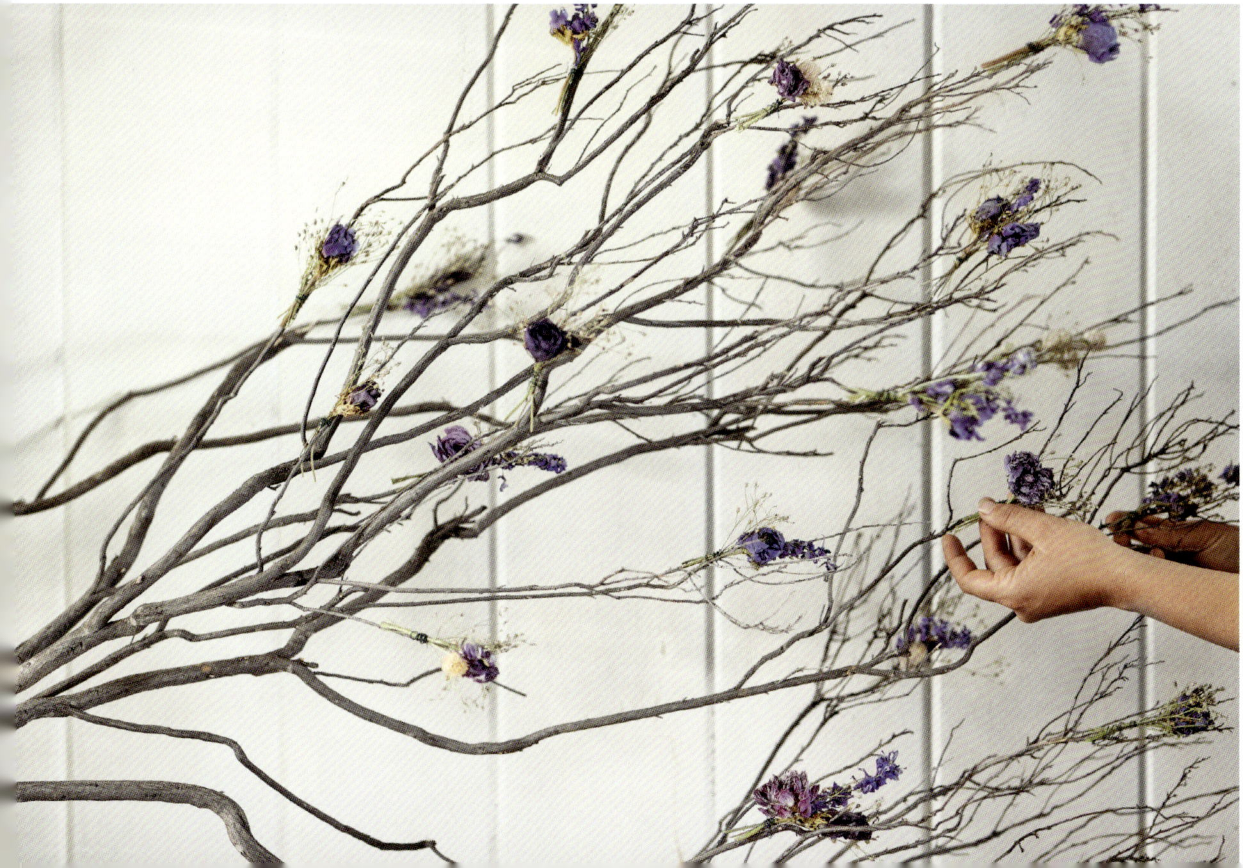

Everything and Anything Mixed Bouquets

MATERIALS

15 to 30 long-stemmed (around 1 foot in length) dried flowers per bouquet

5 to 10 filler and foliage, such as plume celosia

5 to 10 focal flowers, such as craspedia, larkspur, globe amaranth, and black-eyed Susan

5 to 10 grains and grasses, such as millet

Rubber band or twine

Tissue paper

Kraft paper or paper flowers

Ribbon (optional)

THERE IS NOTHING quite like being on the receiving end of a bouquet of flowers. The dried flower bouquets at Full Belly Farm are some of our most popular items at the farmers' market and that we sell to our wholesale accounts. We reserve the flowers with the longest stems for our bouquets. Preparing our dried flower bouquets is much like making our fresh mixed bouquets. We use contrasting tones to make each flower and color pop. Blue larkspur, yellow craspedia, and orange marigolds get nestled next to green safflower, dried sunflowers, and a few sprigs of lavender or sweet Annie to make the whole bouquet smell incredible. These bouquets are the perfect sustainable gift for the holiday season—and the best part is the recipient doesn't need to worry about keeping them in water and can keep them for years to come.

FOR THIS DESIGN

1. Lay the flowers on a table, with stems facing you. Make sure the stems are clean of any unnecessary foliage or flowers that may make the bunch bulky in your hands.

2. Begin to assemble the bouquets by adding foliage and/or fillers first, holding them loosely in between your thumb and pointer finger in your nondominant hand.

3. As you add elements, cross the stems below your fingers. Add focal flowers to the center and front of the bouquet. Loosen your grip as you add more materials.

continues

4.   Complete your design with any grasses to add texture and movement, placing them within the bouquet for additional stem support.

5.   Cut your stems to equal lengths, then wrap the base of your bouquet with a rubber band or twine. To finish it off, wrap the bouquet in tissue paper and kraft paper and secure it with a ribbon or twine.

Fresh Marigold Strands

MATERIALS

50 to 100 fresh marigold heads

26-gauge floral wire

IN THE FALL, the persimmon trees at Full Belly always spark creativity and wonder in me. The dripping fruit hangs low as the leaves begin to turn a brilliant yellow before dropping at the first touch of frost. Recently, I felt inspired to re-create the fruit on a particular persimmon tree that has been at our farm for longer than I have been alive. Something about its expansive branches and the hanging bright-orange hachiyas made me think of the way the marigolds were dripping out in the fields after not being harvested for the past week. I strung marigolds onto wire as I had a beer with my mom after work, chatting as we normally do about upcoming planting days and the many spring flower trials.

Creating flower chains or garlands is a great way to work with fresh flowers and watch them dry in your home or your garden. And because it's a simple activity, it's best done with a small group of friends or loved ones. You can also string other fresh flowers—fresh strawflower or globe amaranth. Both work great and dry wonderfully as well.

FOR THIS DESIGN

1. Cut the wire into various lengths. For this creation, I made twelve garlands in several lengths, starting from 3 feet and ending at 5 feet. But in a home setting, 1 to 3 feet may be best. Add a loop at the bottom of each wire so the marigolds stay on the wire.

2. Push the wire through the soft underside of the marigold and pull it through the center of the flower head.

continues

3. Continue adding flowers until the wires are full, leaving only about 6 inches free of flowers on top to secure the wire to its final resting place.

4. Loop the wire around branches in a tree or on a wall in your house at different heights to create depth.

5. Let the strands hang until dry—but before you do, enjoy these creations by having a picnic or dinner party with the garlands as decoration.

Wreaths

WREATH MAKING AT FULL BELLY FARM

The tradition of making wreaths at Full Belly Farm started when four best friends got together to take a wreath-making class in the late '80s. They hoped this new skill would ignite a passion and help nurture their friendship. My mom and her three friends were in the midst of raising children, growing food and flowers, and keeping homes and husbands, so their friendship had slowly fallen to the wayside. Committing to making wreaths with one another in the fall and winter was a way to recommit themselves to one another and share mothering and farming advice while making money to support their families and farms.

Seeing them gather each fall to make wreaths together, I learned that creating with dried flowers is best done surrounded by community. The women would share advice and their dried-flower collections. They would finish each other's wreaths if someone needed help and offer tips and techniques. Their wreaths would fan out, large and loud, as their voices did the same, chatting to make the time go faster. Years later, we use the same techniques my mom learned with her friends all those years ago. Many of the same women and their children still try to get together each year to spend a day making wreaths in the Wreath Room.

Our creativity is sparked by seeing what other people can make. The conversations are often boisterous and ring out far into the night. This day of design and community gathering is a much-needed boost of inspiration and the motivation that we all need to keep farming and planning for the next year.

As my mom has gotten older and her responsibilities and management role at Full Belly Farm have shifted, she doesn't harvest quite as many flowers as she used to, and the flower team takes on many

of the day-to-day floral tasks. Yet when the busy summer season dies down and her hands itch to make a wreath, she will often join me in the Wreath Room in the afternoon and tell me stories of those early years of making wreaths with her best friends. I am always blown away with the talent she possesses to make immensely creative wreaths. Not only are her wreaths beautiful and unique, but her efficiency is a true testament to a lifetime of skill and muscle memory.

My mother graciously and patiently taught and continues to teach wreath making to the flower farmers and the interns passing through the yearlong program at Full Belly Farm. I sometimes like to imagine the web of artists that my mother has helped to teach and inspire. There are wreath makers across the country and world now who all practiced and learned alongside her in the Wreath Room.

Basic Wreath Technique

MATERIALS

Homemade round

26-gauge floral wire

125 to 200 stems-dried flowers, 60 to 100 filler flowers, 33 to 50 focal flowers, and 33 to 50 grasses and grains

Broom corn

Cockscomb

Feverfew

Marigold

Millet

Plume celosia

Strawflower

Sunflowers

I STILL REMEMBER learning to make my first wreath when I was five or six. I was delighted that my little hands could create something so inventive. Once you learn the basic techniques, you can experiment with many different bases, decorative touches, and flower combinations. While the fundamentals will stay the same, you can test the boundaries of what this craft holds and continue to find your unique voice as a designer. Just like any skill, making wreaths takes practice and patience but, above all, they require a passion to continue to discover new depths of the art form. The techniques used and taught below will be the foundation for every wreath you see in this next section. Once you learn the basics, it's up to you what you choose to create next!

CHOOSING A BASE

Three common bases for wreaths are pictured here.

1. DOUBLE-WIRE BASE

These bases are great for greenery wreaths or wreaths that have thicker, heavier flowers or foliage where stems may need more support. Some double-wire bases have clamps that you press down as you progress, or they can be used on wreath machines that use a pedal to clamp the wire in place.

2. GRAPEVINE BASE

A grapevine base can easily be handmade with slightly green grapevines before being pruned back in the fall. They are also easy to purchase premade in a variety of sizes. Vine wreath bases are perfect for half or quarter wreaths, where part of the wreath base will be exposed.

3. HOMEMADE WIRE ROUND

These wire bases are what we use at Full Belly Farm because they allow us to make many different shapes and sizes, and they cut down on the cost of materials. Each of our bases is made with 14-gauge baling wire, are 5 to 8 inches in diameter, and are wrapped twice around for additional support. You can also use a metal coat hanger in a pinch.

CHOOSING
YOUR FLOWERS

Select ingredients that help support and fill the three main sections of your wreath: the outermost side, the top, and the inside. Each flower type will play an important role in the structure, texture, and complexity of your wreath. Grains and grasses make excellent sides as they often naturally fan out. They also add great flow and movement to wreaths. Fillers and foliage will make up the majority of the wreath and help fill in holes. Their fullness will help keep the focal flowers front and center—making them the stars of the show.

We generally work with five or six ingredients in our wreaths. Start with two or three bunches of each of those five or six ingredients, approximately 150 stems total. You can adjust the amount, color, and variety of each flower to get a myriad of different outcomes. Just be sure to include about 50 percent filler, 25 percent focal flowers, and 25 percent grains and grasses to ensure that the wreath feels full and there will not be any holes in your design. If you have less of something, make sure you use that ingredient sparingly throughout your design as it's difficult to create a full wreath if you run out of a material halfway through. If the flowers feel brittle or break easily as you start your design, make sure your ingredients are properly rehydrated (see Brittle Flowers, page 42) to make them easier to work with.

STARTING OUT

1.   Cut the flowers down to 4 to 5 inches below their heads. Some of the grains, grasses, foliage, and fillers may need to be split apart or broken down into smaller pieces. Anything too large can weigh down the wreath or make the base too bulky. Strip any foliage off the stems of the flowers to keep the stems clean. Keep the process smooth and straightforward by organizing the flowers into small piles by type in front of the wire round.

2. Attach 26-gauge floral wire securely anywhere between one and three o'clock on the wreath round by wrapping it multiple times around the same spot. This will be the wire that holds every flower on to the base. Never cut the wire until the very end. Using the same continuous strand will ensure all the flowers stay secure, bundled together, and the wire stays tight. If your wire breaks or gets cut accidentally, reattach the wire to the round before you continue.

3. Gather a small collection of flowers and place them on top of the wire round, right above where the 26-gauge wire was first attached. All the stems should be oriented clockwise around the base. The flowers should sit on the top or outside of the wire base, never underneath. This will ensure that the wreath will rest flat on the wall once completed.

4. Attach the first small bundle of flowers by securely wrapping the wire around the flowers and the wreath round three times, about 1 inch below the base of the flower heads. Once those are in place, place a few more flowers snugly below the necks of the flowers previously attached. The stems should all be going in the same direction. Secure those flowers to the wreath base, slightly lower than the previous spot, by wrapping with the floral wire three times. I find it helpful to use my nondominant hand to hold the flowers in place with my index finger and thumb while looping the wire with my dominant hand.

GOING ROUND

5. Continue this process *slowly*, attaching more flowers below the previous flower heads and keeping the stems together on the top or outside of the wreath round. If the flower stems get too unruly or long, clip them to about 3 inches from the spot you last attached them. The previous stems should help create the base that your next flowers sit on.

6. Focus your attention on building the outside, top, and inside of the wreath. The outside should fan out, creating the majority of the mass of the wreath, while the top should mostly be to help cover your previous work and any exposed wires and to begin building your pattern. The inside should be where you place the smallest number of flowers. Those should be added sparingly and only to keep the wires hidden and circular shape even.

7. Remember that you will work your way around the wire frame just once. So the wreath should look full and lush the first time through. Do not forget to fill in holes with filler and foliage and create a pattern with the focal flowers by placing

them in every bundle or every other bundle you attach.

8. Turn the wreath slowly as you move forward. It's best to try to keep your work always at twelve or three o'clock so you know when it's time to move forward and you don't get stuck in one spot. There will be moments in the middle when you will want the process to go faster; but just remember to take your time, breathe, and enjoy the process.

FINISHING UP

9. The trickiest bit of any wreath is completing the circle. Many florists choose to add a ribbon or bow once they return to their starting point, saving themselves the headache of finishing the wreath in full. But once you know how to finish a wreath, there is no better feeling than seeing the circle of flowers completed. When the stems of your current bundle of flowers begin to touch the point where you started the wreath, tuck the stems behind the beginning of the wreath instead of going over the top of the wire, being careful not to cover up the start of the wreath with the current stems. Once you have filled the last space, cut the floral wire about 3 feet long and, instead of going over the stems with the wire, thread it through your starting point to attach the last batch of stems without crushing them or going over the top. Finish it off by tying the wire to the wreath round so it's secure and cutting any excess wire that remains.

10. At this point, you can finesse your design to your liking. Turn the wreath slowly to survey it for holes or any missing spots in your pattern. You may notice a natural top or bottom to the wreath, where the flowers feel fuller. Remove any extraneous flowers with a pair of clippers. Use a small amount of floral glue or hot glue to put any remaining touches in place. Finally, hang the wreath on a wall and proudly stand back and admire your hard work.

A NOTE ON MAKING YOUR DRIED WREATH LAST

❃

Wreaths can last indefinitely if handled and displayed correctly. Hang it indoors where it's not in direct sunlight, not too dry, and not exposed to moisture in the air. If your wreath gets dusty or has cobwebs, a feather duster works perfectly to remove dross without damage. We have happy customers who purchased our wreaths more than a decade ago and claim the wreaths look as good as they did on the day they were bought.

THE ARTISTS
AT FULL BELLY FARM

There is a sound that I associate with wreath season at Full Belly Farm. It's the gentle noise of twelve hands moving slowly, the rustling of flowers as small bundles are attached to wire rounds, the floral wire clacking on the table as wreaths are slowly turned forward. The Flower Ladies work from October to the middle of December making wreaths. Isabel, Fidelia, Antolina, Sonia, Maria, and I make wreaths from eight in the morning until four in the afternoon. Each one is created with consideration and creativity.

We sell our wreaths at farmers' markets, to stores and restaurants, and through our CSA boxes. A tag with the artist's name is placed on every wreath so buyers know that each one is a labor of love. Churning out wreaths at a large scale can be challenging. Each one takes about an hour and a half

from start to finish, and creating new and unique patterns and designs means constantly being innovative. While we could probably create a more mainstream and profit-driven way to market and make our wreaths, part of the appeal to us and to our customers is that the designs are small-batch and true works of art that will hang in someone's home for years.

Each wreath maker has their own style. Isabel has been making wreaths the longest out of all of us. Her creations are wild and filled with grains and grasses that look like prairie fields. Antolina is unmatched in her patterns—each one a puzzle of beautiful designs and explosions of color. Fidelia always makes hers sleek and colorful and the perfect size and shape for our wreath boxes. Maria's color palettes make us all envious, mixes of blues and purples, browns and yellows, in compelling combinations. Sonia is no-nonsense and practical, making wreaths the fastest and choosing patterns and flowers that are bold and contrasting. We make a unique team, each with our own talents. and providing a wide range of wreaths that all vary, based not just on the flowers we grow but on the personalities of the artists.

FIDELIA

MARIA

ANTOLINA

SONIA

HANNAH

ISABEL

Aromatic Lavender Wreath

MATERIALS

*Between
200 and 300 stems*

IN LATE MAY, we begin harvesting lavender to dry. The trick is to capture it right before it begins to fully bloom, which, when you have two 600-foot rows of lavender all blooming at the same time, can be quite hard. If you wait too long, you will end up with flowers that are dull and shed and break easily. As the sun peeks over the hills and the lavender turns from a rich gray to a fluorescent purple in the light, we begin our late-spring ritual. We clip giant handfuls, rubber-band the bottoms, and toss them into the furrow, where they lie until we break for the day and carry our armloads into the Wreath Room to hang. The bunches don't need to be perfect. Efficiency and rhythm are key. The bees keep us company as we harvest. Swarming and sipping, not at all concerned with our hands or fast-moving clippers. But I do like to wait until the bees have had a moment to wake up. The only time I have been stung is when I have startled a sleeping bee.

Once the lavender is harvested and hung, the Wreath Room turns into a den of fallen petals and hanging blue bunches that smell earthy and pungent. We often gather the lavender petals into bowls or baskets, holding them for winter, when we add them to satchels for holiday stocking stuffers or mix them into flower confetti for wedding celebrations. When creating lavender wreaths, you should expect a natural amount of shedding from the flowers. Collect the aromatic buds that fall for later projects. To make the process of lavender wreaths go faster, and make the final product look more full, I generally bundle about 10 stems of lavender together in my hand every time I attach more to the wire round.

FOR THIS DESIGN

❀

150 stems lavender

15 stems strawflower

50 stems amaranth leaves

15 stems ranunculus

50 stems plume celosia

FOR THIS DESIGN

❋

100 stems lavender

*50 to 75 stems globe
amaranth*

50 stems plume celosia

50 stems nigella

*15 to 20 stems
strawflower*

Naturally Neutral

MATERIALS

*Between
50 and 100 stems
per wreath*

FOR THIS DESIGN

Amaranth

Ammobium

Broom corn

Feathered celosia

Flax

Globe amaranth

Millet

Nigella

Poppy pods

Wheat

ONCE WE UNPACK OUR BOXES of flowers in the fall and prepare for the months of wreath making that lie ahead, my first inclination is to overdesign. With so many beautiful flowers to choose from, all enticing me to use them in my next creation, the combinations and color palettes seem endless and exciting. Yet there is something peaceful and modern about limiting what you use in a wreath and allowing the few ingredients and unique textures to speak for themselves.

For this series of mini wreaths, I kept the color palette simple, opting to let the different grains and pods shine. I then turned my attention to the flowers. I chose things that had a variety of unique shapes and textures and were all naturally neutral ingredients.

I made each wreath round, smaller than our normal-size base, ranging from 2 to 5 inches in diameter. I added a simple ribbon to a few of the final wreaths and kept the stems long at the end of the others rather than completing the round fully to give the wreath a more natural look. This is a great way to use products if you have a limited supply, want a wall full of wreaths that match, or want a more modern look.

Monochromatic Moments

MATERIALS

*Between
120 and 200 stems*

MANY DRIED FLOWERS have an incredible knack for acting as unique blending tones between bright colors. The fading that happens naturally in the drying process, as well as the variety of pigments that show up in many of the flowers that dry well, lend themselves to creating a beautiful natural scale of monochromatic hues. These next wreaths show just how many colorways are possible while sticking to predominantely one color per wreath. Monochromatic designs bring a sense of cohesion and clarity and can help bring the attention to the textures of the designs, rather than the color. I have found that monochromatic wreaths look more modern and bold—and are a unique take on an age-old tradition. Try creating a red-only wreath to hang on your front door or create a yellow-hued wreath to hang on your blue bedroom wall.

FOR THIS DESIGN

YELLOW	BLUE	PINKS AND RED
Craspedia	*Globe amaranth*	*Cockscomb*
Feverfew	*Larkspur*	*Globe amaranth*
Flax	*Lavender*	*Pennycress*
Nigella	*Poppy pods*	*Plume celosia*
Ranunculus	*Ranunculus*	*Ranunculus*
	Statice	*Safflower*
	Strawflower	*Strawflower*
	Wheat	*Wild mustard*